HACK THE EXPERIENCE

Before you start to read this book, take this moment to think about making a donation to punctum books, an independent non-profit press,

@ https://punctumbooks.com/support/

If you're reading the e-book, you can click on the image below to go directly to our donations site. Any amount, no matter the size, is appreciated and will help us to keep our ship of fools afloat. Contributions from dedicated readers will also help us to keep our commons open and to cultivate new work that can't find a welcoming port elsewhere. Our adventure is not possible without your support.
Vive la open-access.

Fig. 1. Hieronymus Bosch, *Ship of Fools* (1490–1500)

First published in 2018 by Brainstorm Books
An imprint of punctum books, Earth, Milky Way
https://punctumbooks.com

ISBN-13: 978-1-947447-65-3 (print)
ISBN-13: 978-1-947447-66-0 (ePDF)

LCCN: 2018945514
Library of Congress Cataloging Data is available from the Library of Congress

Book design: Vincent W.J. van Gerven Oei

HACK THE EXPERIENCE

New Tools for Artists from Cognitive Science

RYAN DEWEY

Brainstorm Books
Santa Barbara, California

Contents

0

Introduction

Artists and scientists do the same thing: they observe something, analyze their observations, and present their findings in a way that untrained people can see *with relative ease* what the artist or scientist worked to see in their own analyses.

The core argument of this book centers on the premise that art is a form of cognitive engineering and that the physical environment (or objects in the physical environment) can be shaped using empirically validated models from cognitive science to maximize emotional and sensory experience. Like the title suggests, this is a kind of hacking. As the book unfolds, this kind of hacking will take various environmental resources and interventions and apply them to different conditions (emotions, senses, general experience, etc.), in order to design integrated experiences that help awaken your audience in new ways.

Cognitive engineering blends the role of artist and scientist into a process of building experiences that evoke responses of some kind. This book will give you a toolkit for planning experiences and events of all sorts and will guide you through the process of bringing elements of cognitive engineering into your creative practice. After reading this short book, your work will help your audience connect with your work in more meaningful and memorable ways. **Cognitive engineering in this sense provides a way to promote inquiry, and to make people curious about the world. The experiences you design will inform the way that people live in the world and make meaning in their day-to-day life.**

Artists naturally evoke responses from their viewing audience, such as the visceral responses of disgust, of joy, and also cognitive responses of indifference, bewilderment, confusion, hatred, or suspicion. This in part comes from the content of the work, partly from the context, partly from the execution, partly from the past experiences and openness of the audience, and partly from scores of other factors. All art evokes responses. "My kid could do that" is a response, and, like it or not, it is a valid response that was evoked by the work. The goal of this book is to help you shape the viewers' experience to leverage basic cognition and attention patterns to push people toward a set of responses you want to evoke. For instance, if you want to create a feeling of remoteness in the middle of a city, the tools in this book will help you evoke that feeling by showing you which perceptual elements you need to shape in the physical space and how.

Disciplines like new media studies, neuroscience, biosemiotics, and data studies already define significant corners of the contemporary art world, so there is a context for creating experiences that make sense out of the body and the brain. This book contributes to that effort by putting ideas together that can move your existing creative practice up onto new levels of experiential grounding and perceptual relevance. It is designed for people who are actively working on projects, and it intervenes into your creative process like a critic and a consultant, guiding you with new insights into your process of creation. This book is dense, like a bullion cube of concepts and methods. It is good advice to read this book at least twice, and to read it while working on specific projects. Also, as soon as you learn a concept from this book, try to teach it to someone, explaining it in your own words so as to further solidify the concept in your own memory.

This guide is also useful to people working in museums and galleries. In the turn toward experience in museum interpretation, this guide makes sense of the ways that audiences engage with stories told through juxtaposition, organization, and collection, helping to bring an outside eye to curatorial practice. Creating experiences from the ground up in a requirements-driven process lets experience designers play around with a variety of methods to engage the public on emotional, visceral, and cognitive levels. It makes sense then to have some understanding of the way general cognition affects experience to better exhibit certain works, and to design exhibitions that better expose the relationship between cognition and experience.

Outside of the gallery in the applied arts (like design), engineering response-evoking user experiences is also a large part of brand development, and this simple guide can even serve as a playbook for designing transformational experiences that add perceptually relevant texture to brand narratives. While the focus of this book is art installation, the presented ideas equally apply to building stories and worlds, whether for scripts, storyboards, or game design.

What constitutes "hacking"? Hacking in this book is an intervention that organizes the physical world in a way that leverages natural cognitive structures to evoke some kind of response in the viewer. It's "hacking" because it is a kind of short-cut into a mental state through various openings in the body-mind-experience system.

What exactly does "experience" mean? This word has two senses in focus here. First, experience refers to the fabric of moments in general. Second, and more specifically, experience refers to your neural integration of stimuli in the environment to create your perception of a moment in time. This first sense is the perceived state: the experience of the participant. The second sense refers to the atmospheric and event-like nature of experience: experience as a moment in time in some designed environment, installation, or intervention (i.e., the moment created by the artist or designer), and how that is comprehended by the viewer-participant. In other words, the first sense is your perspective and the second sense is the context in which you have your perspective—your experience (first sense) of an experience (second sense).

This second sense of experiences includes things like performances, gallery exhibits, parties, festivals, events, happenings, immersive theater, installations, new media art, pop-up spaces, stable architecture, urban fabric, cultural narratives, and virtual and physical worlds. They can even be small-scale, private events like brushing your teeth, or drinking a glass of water, but they can also be massive in scale, like the string of activities involved in rebranding post-industrial cities. *Experience can be shaped as it scales up or down.*

Experiences can easily be expanded to include theme park design, museums, restaurants, brand encounters, advertising campaigns, and film. An experience can take place in a single location (such as a standard event or party) or it can extend over several locations (such as a hike or even a tourist walk). An experience can have a single audience or a very large audience. It can have one or many goals, it can consist of one activity or many activities, it can occur once and be over or it can extend over a long period of time, sequentially in some predetermined order or *ad hoc* as determined by happenstance or by the random paths of audience encounter. There are no limits to the characteristics of experience, except for the requirement that it has to happen *somewhere* to *someone* at *some time* and it has to do *something* to that person, such as evoke a response.

Experiences in the first sense happen whether or not experiences in the second sense exist at all. **But experiences in the second sense are designed to intentionally influence first-sense experience.** Second-sense experiences center around *intentionality* of design, *intending* something to mean something for someone else. It is a kind of language through which the artist communicates something to the audience. The artist *means* something for the audience. As an artist you already know how to communicate to your audience through the traditional channels of form and subject and content. This book works to uncover how structural elements of cognition occur pervasively in daily life and how they can be harnessed in immersive, spatial approaches to art.

Cognitive engineering helps orchestrate any kind of experience: dinner for two, a site-specific installation, an exhibit, a building, an event, a relationship, an identity, a life story, or a brand narrative. This handbook sets out to show you how to set up scenarios for your audience to engage with experiences in ways that encourage them to oscillate between the roles of participant and spectator. Follow this guide to help your audience *think through action*. Your audience will experience your work and ideally have a richer experience of the world at large because of how you bring in awareness, the senses, and basic narrative forms to create compelling installations.

In his work on cognition and aesthetics, neuroscientist Merlin Donald (2009) defines art as a form of cognitive engineering and argues that the main goal of art is to evoke responses of some kind in the viewer. The production of art is a process of changing the way people see some element of experience by engaging people through their experience of a specific work of art. Artists have always engineered viewer experiences, but new tools from cognitive science (like conceptual metaphor, blending, and cognitive simulation) enable artists to refine this process through the use of cognitive models (of attention and embodiment) as foundational elements in the same way that they use materials and the range of techniques in traditional and contemporary art making.

Taking the mind as the target audience of an artwork, it makes sense for artists to use models that come from the empirical findings of the sciences of the mind to enrich and underpin an art practice.

Pay attention to this statement, it will shape the way you make sense out of the models presented in this book: **Understanding the basic patterns of human attention unlocks the door to how you shape the viewer's experience of an artwork. Everything is about making the most use of attention.**

The book is divided into four sections:

Section 1: Hacking & Engineering Experiences

This section looks at the process of hacking cognition, the identification of backdoors for hacking, the role of design, understanding the human attention system in order to direct it, design armatures, and practices of thinking about design problems in experience hacking. This section presents some of the ideas that act as underlying skeletal structures in designing experiences.

Section 2: Toolbox of Cognitive Tools

This section has seven tools that give you important background information on different elements of experience and what they are useful for in designing experiences. These are topical sections that get you thinking about the tools that are available to you as you design experiences. This section presents some of the musculature of everyday experience and outlines ways that these experience elements can be combined, modified, and experimented with in the process of designing physical interventions in the design space.

Section 3: Stories & Paths

This section ties together the musculature from Section 2 and the skeleton from Section 1 and grounds experience design in storytelling processes along paths and at intervention nodes along these paths.

Section 4: Documentation & Interpretation

The types of experiences that you will build with these tools will be hard to describe and difficult to sell to institutions. This section presents a robust model of documentation for planning, archiving, and reproducing your work and then lays out a plan for working with museums and galleries to help them understand how your work fits with, and also serves, their operating goals.

When you are done reading this small handbook you will know enough to be able to complete the following steps for building an experience that is customized to your artistic vision. You will be able to:

- think about your installation as a story;
- provide your audience with a framework for inquiry into the sensory world;
- pair two or more sensory systems, viewpoints, conceptual models, physical systems, or elements of embodiment into sensory metaphors;
- couple those experiential pairs with mental, emotional, or physical information;
- introduce a pattern of when and how specific sensory systems are active or activated;
- build indeterminacy of participation into your experience to give your audience the ability to control part of their experience;
- create engagement points for triggers, feedback, openness, and bodily responses;
- shape experience in simple ways to create new effects in your installations;
- capture, focus, and direct visitor attention through space, time, and information;
- use distraction appropriately by eliminating or creating it as it fits your story/goal;
- document your work as thoroughly as possible/necessary; and
- offer an experience that is designed to be memorable.

Manage Your Expectations about This Book (What it Is and Isn't)

This book is not designed to create an art project for you, but rather gives you elements that you can work into your own existing practice. This book doesn't specify whether your art project should use digital technology or chemistry or plastic. It doesn't specify the mechanisms you should use to produce your art. It does not spell out, start to finish, how to produce any kind of work. **What it does do is give you doors into topics that can radically shape the work you produce.** It will help you situate your work in a broader discourse of the mind (specifically, how a unified mind and body engages the environment), and it will help you think about complex work in systematic ways. It won't hold your hand, but it will point you in the right direction. You still have to do the work to make these approaches fit your practice. This is a strength of the book. The armatures and tools in this book can be effectively applied to any subject, in any form, and with any content. It is up to you to make the connection between the ideas presented in this book and the process you use in your practice.

This is not a science book. It is not meant to lay out all of the relevant research in cognition. I've tried to keep citations to a minimum, and instead, point readers to texts that address specific concepts in cognitive science. This is because a book like this can't integrate all of the research in cognition. Cognitive scientists might find this book boring. They might feel that I don't do justice to some particular concept in the relevant research. That is not the point of this book. I want this book to be a bridge between disciplines. I hope that it will also inspire other scientists to expand on the concepts presented in this book.

This book also does not have allegiance to any particular theoretical framework of contemporary art. Besides, you probably know much more than I do about contemporary art—you may know more examples than I do, you may have a more nuanced grasp of the philosophical issues of art. This book is not a philosophy book either. It does not have a single reference to some of the more trendy philosophers that many contemporary art schools have their BFAs and MFAs read. I believe the dialogue between artists and scientists should go both ways. I hope that it will also inspire other artists to expand on the concepts presented in this book.

This is not a book about technology, and it is not necessarily about game design, nor about coding. You can apply this book to those domains because this is a book about concepts that apply to people and their experience of the world. No matter what your

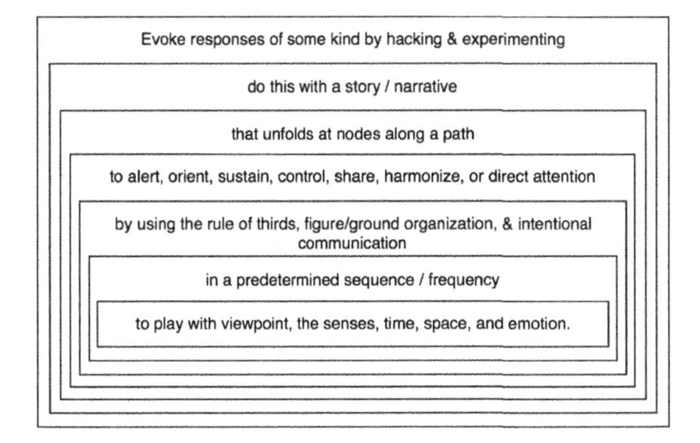

background, **if your work involves *language, thought, perception, or behavior,* then this book will have something to offer.** If not directly informing your work by supporting and confirming your views, then this book can indirectly inform your work by challenging your views.

Also, this book is not about how to address science as your subject and content in your artwork. Yes, you can apply the principles here to your practice if you happen to create science-themed art, but these same principles can apply just as readily to art that is non-scientific in terms of theme or content. This book views experience, *the senses, emotions, memories,* and *the body* as mediums for content.

This is not a book outlining all of the methods of narrative development. It uses a basic notion of narrative to make it accessible to a broader audience, but many of the concepts can apply equally well to other models of more complex narrative structure.

This is a book about you using your art practice (whatever that looks like) as a form of cognitive engineering. You are the expert on what you do; this book simply provides a new framework for thinking about your work.

This book has a very simple message. Experiences that use story engage an audience more effectively because the story frames how people enter and move through the experience. Story helps offset the personal idiosyncrasies of your audience members because you construct the story as its own system and your experience helps people live in that system for a moment. It establishes a common ground benchmark to equalize your otherwise diverse audience in some way. What follows in this book is an approach to have cognitive tools support your storytelling so that you can leverage some basic human systems like *perception, emotion,* and *attention* to help people join in your story world as *participants* and *observers*.

This book is not meant to give a definition of what art is from a cognitive science perspective, and it is not a work of neuroaesthetics. I agree with John Hyman's critique of neuroaesthetics for drawing broad and fast conclusions about art in general and assert, with him, that "we should be pluralists about artistic value" (Hyman 2010, 260) rather than trying to define all of art according to neurologic generalizations. This book does not attempt to explain the mystery of art, but instead is meant to provide one set of different tools to artists that enables them in the production of art to make use of the attention patterns of viewers in new ways as they experience the artistic "product," in whichever way the artist defines that artistic product. This is one reason why this book does not approach empirical research in a dogmatic quasi-idolatrous way, favoring instead to lay out the ideas to allow you to pick and choose which elements most effectively serve your artistic program. Instead of prescribing rules, I want to present some concepts which I see as related and relevant to the production of art installations and let you tweak those concepts as you see fit. As a scientist, I am unashamed to admit that science doesn't know everything, nor will it ever, in my view, but maybe science can point out some useful clues to move us as artists in a good direction for new inquiry.

Before moving to Chapter 1, take a minute to read the goals of this book as a sentence (Figure 1) to get a peek at how the book will develop.

Figure 1. What to Do.

1

Breaking Down Experiences to Find Backdoors for Hacking

Think back over your entire life. Can you remember everything? What things do you remember? Some experiences are more memorable than others. As an artist you produce works that in most cases you want your audience to remember. This means you probably want to make your work stand out. But what makes anything stand out?

Let's start by looking at the different parts of an experience, and then let's reverse-engineer an experience in order to see how aspects of the experience combine to structure that experience. Dissecting an experience will help in identifying how we can hack into people's experience to produce enhanced visceral effects.

Experience is more like a fabric than a single thread. It is multiple linear threads woven together to form a meshwork of individual threads (a fabric) that is something altogether new. Fabric is flexible and multidimensional, whereas thread is a linear and unidimensional element. You can think of the different elements of your experience (like the senses, time, spatial context, emotions, memories, etc.) as different threads, and when those threads come together, they create the fabric of experience in general. Designing an experience requires playing with the threads of experience to organize a pattern that you want to structure the moment of experience for the visitor.

We often dissect things that we want to understand, and experiences are no different. Taking apart the fabric of experience means that we separate experience into the different threads that make up experience. But what makes up an experience? Let's look first at everyday experiences as a way to learn how to hack into the deeper structures of experience.

As a designer of experiences, you are engaging in the process of building culture. It makes sense then to look to cultural researchers to see the different ways they look at experiences. Ethnographers learn about cultures by participating in and observing different cultural practices. They use a variety of systematic approaches to understanding cultural practices and their goal is to describe a culture from both the insider and outsider perspective. One approach that is particularly useful to artists and experience designers is the approach called *descriptive observation.* An ethnographer doing descriptive observation goes into a culture and asks descriptive questions about cultural experiences. James Spradley, in his book *Participant Observation* (2016), outlines a series of nine dimensions for experience: *space, object, act, activity, event, time, actor, goal,* and *feeling*. Together, these dimensions add up to create the fabric of an experience (from the perspective of ethnography). In Spradley's approach, all of the dimensions help the researcher build questions to ask about the cultural experience (Table 1). This approach looks at the **spatial**, **temporal** (*spatio-temporal*), and **functional** relationships within a cultural experience, all organized around understanding the members of the culture by accounting for the knowledge it takes to be considered a member of that culture. In that sense, it is an **agent-activity focused approach.**

	SPACE	OBJECT	ACT	ACTIVITY	EVENT	TIME	ACTOR	GOAL	FEELING
SPACE	**Can you describe in detail all the *places*?**	What are all the ways space is organized by objects?	What are all the ways space is organized by acts?	What are all the ways space is organized by activities?	What are all the ways space is organized by events?	What spatial changes occur over time?	What are all the ways space is used by actors?	What are all the ways space is related to goals?	What places are associated with feelings?
OBJECT	Where are objects located?	**Can you describe in detail all the *objects*?**	What are all the ways objects are used in acts?	What are all the ways objects are used in activities?	What are all the ways objects are used in events?	How are objects used at different times?	What are all the ways objects are used by actors?	How are objects used in seeking goals?	What are all the ways objects evoke feelings?
ACT	Where do acts occur?	How do acts incorporate the use of objects?	**Can you describe in detail all the *acts*?**	How are acts a part of activities?	How are acts a part of events?	How do acts vary over time?	What are the ways acts are performed by actors?	What are all the ways acts are related to goals?	What are all the ways acts are linked to feelings?
ACTIVITY	What are all the places activities occur?	What are all the ways activities incorporate objects?	What are all the ways activities incorporate acts?	**Can you describe in detail all the *activities*?**	What are all the ways activities are part of events?	How do activities vary at different times?	What are all the ways activities involve actors?	What are all the ways activities involve goals?	How do activities involve feelings?
EVENT	What are all the places events occur?	What are all the ways events incorporate objects?	What are all the ways events incorporate acts?	What are all the ways events incorporate activities?	**Can you describe in detail all the *events*?**	How do events occur over time? Is there any sequencing?	How do events involve various actors?	How are events related to goals?	How do events involve feelings?
TIME	Where do time periods occur?	What are all the ways time affects objects?	How do acts fall into time periods?	How do activities fall into time periods?	How do events fall into time periods?	**Can you describe in detail all the *time periods*?**	When are all the times actors are "on stage"?	How are goals related to time periods?	When are feelings evoked?
ACTOR	Where do actors place themselves?	What are all the ways actors use objects?	What are all the ways actors use acts?	How are actors involved in activities?	How are actors involved in events?	How do actors change over time or at different times?	**Can you describe in detail all the *actors*?**	Which actors are linked to which goals?	What are the feelings experienced by actors?
GOAL	Where are goals sought and achieved?	What are all the ways goals involve use of objects?	What are all the ways goals involve acts?	What activities are goal seeking or linked to goals?	What are all the ways events are linked to goals?	Which goals are scheduled for which times?	How do the various goals affect the various actors?	**Can you describe in detail all the *goals*?**	What are all the ways goals evoke feelings?
FEELING	Where do the various feeling states occur?	What feelings lead to the use of what objects?	What are all the ways feelings affect acts?	What are all the ways feelings affect activities?	What are all the ways feelings affect events?	How are feelings related to various time periods?	What are all the ways feelings involve actors?	What are the ways feelings influence goals?	**Can you describe in detail all the *feelings*?**

Table 1. Spradley's *Descriptive Question Matrix* (2016). Used with permission, courtesy of Waveland Press.

There are other ways to break up the experience. You might break it up into **narrative structure**, looking at the way that a story is told in and through the experience to see how the path of a story ties everything together. You might also break up an experience into the different **cognitive elements** that are at play, such as attention, meaning, or communication. Or you might break it up by the **visceral elements**, such as the five different sensory channels (sound, taste, touch, smell, and sight), and the **emotional responses** evoked by the experience. By looking at these dimensions you get a sense of the way that an experience is made up of many different threads. Any and all of these threads are back doors where you can begin your hacking.

Each approach to understanding experience takes a focus, whether that focus is the actions people undertake (agent-activity focused), the stories they tell (narrative), the way they think (cognitive), the way they sense (visceral), or the way they feel (emotional). These different approaches are outlined in more depth throughout this book. There are additional approaches that look at the mechanics and technologies employed in an experience, and some of these are explained in Chapter 9 on methods of documentation. Because this book is primarily focused on hacking the experience of people, more emphasis is placed on the approaches to experience that emphasize people rather than approaches that emphasize technology. This book is technologically agnostic because keeping pace with technological changes is impossible. No matter what type of technology your experience uses, whether it is digital or analog, every technological intervention should be body-centric: they should access what Nathaniel Stearn (2013) describes as the "*implicit body*" because the body is our primary tool for thinking and feeling and experiencing the world.

You can use all of these people-centric approaches to categorizing the pieces-parts of experience in a simple, three-step format:

1. Break the experience into parts;
2. see how one part relates to the other parts; and
3. see how each part provides structure to the experience.

Then, with a goal toward designing an experience that hacks human experience, add another step:

4. See which parts are better back doors for hacking (which have biggest impact).

You can determine which parts are better back doors for hacking by determining the intensity of the consequences of either removing, muffling, or amplifying a particular part as identified in Steps 1–3. Like designing experiments, a good amount of guessing and hypothesizing is involved. Start by testing out the simple presence and absence of a particular part and then experiment with different intensity thresholds of presence and absence. Test out how present or how absent something needs to be in order to achieve the effect. "Does sight need to be present in order to experience this work? If yes, can I enhance or decrease vision with lighting to play with the effect?" Test out whether you should present parts in isolation or whether the parts should be integrated in new combinations. "Should what people see match what they hear? Can I correlate what people hear with what they smell? Can I give participants a spectator view of themselves? Can I give a spectator the immersive view of a participant?" It might be that some parts will distract from your goals in the experience you are designing, or it might be that missing parts cause the audience to fill in the missing details for themselves (ideally with inferences that are tightly connected with their emotional investment in the experience). This takes you to Step 5:

5. Harness those back doors into your designed experience.

By identifying the parts that build the fabric of the experience and which parts are good for hacking, you have already begun to identify where your content fits in and how your content can put a skin on the skeleton of the experience structure. Now you need to put those back doors into some kind of overarching structure in order to build a coherent experience. Using a narrative arc (just like with any story) is a good place to start. Turn the back doors into stopping points along the narrative arc. Later we will see how to relate your story structure into a spatial layout in the gallery.

This was a dense, if brief, chapter. It would be wise to take a break for a couple of days to look at moments you encounter through these new lenses.

2

Control Attention

All art involves control or direction of attention.

The first step to hacking experience is to understand how basic human attention works. Once you understand the basics of attention, you can modulate attention in your designed experience and leverage attention to evoke the responses you want in your audience. Understanding basic attention answers the question—*why does anything stand out?*—and it gives you the tools to make things stand out or blend into the background in ways motivated by your designed experience.

Humans recognize intentionality from a very young age. From the time that an infant is around nine months old they recognize intentionality in movement (Tomasello 2003). Try this the next time you see a baby: roll a ball in front of an eight-month-old baby and they will find it amusing, nine-month-old children find this boring. Balls that roll smoothly don't exhibit intentionality on the part of the ball and the nine-month-old child recognizes that the ball is not moving on its own because they know that you have set the ball in motion. They sense that the path is too regular and the movement is too uniform. But if that ball is designed to wobble or to follow an irregular path, the situation is different. A nine-month-old child will see variable motion as an indicator of intentionality and will be more interested in how the ball moves. This also works with cats, however old, and of course, it works on adult humans too.

The ability to recognize intentionality develops as the child learns to communicate, and eventually helps the child to incorporate intention into its own communication. Communication critically depends on the mutual recognition of intentionality between two parties. Another developmental skill that we acquire as infants is **joint-attention**. Joint-attention uses intentionality as a building block for coordination between individuals.

One way coordinated joint-attention happens is through simple *contact* and *following*. Mothers and babies do this early in life when the mother looks at something in order to get the baby to follow her gaze to look at the same thing. It is a way of pointing with our gaze and it is a basic way of communicating meaning between two people (Figure 2). Film directors incorporate this concept when they cut from one frame to another where the first frame portrays an actor looking in a particular direction and the following frame displays what the actor is looking at. It creates the effect of joint-attention because it communicates to the viewer of the film what the actor is looking at (Tobin and Oakley 2010; Brown and Dewey 2014). This works in film because it mimics a basic pattern of attention that we learn in infancy. You can make this work in your art by controlling where people look by coordinating gaze and controlling which views are available in the visual field.

Coordinating gaze through contact and following is one way to "mean something" for another person. What are some other ways that we can mean something for our audience? Alert people's attention to some stimuli. How can you make that stimuli stand out? Inject your new stimuli into a consistent and contrastive background. Here is an easy example from every day life: in a silent room, breaking the silence by speaking is an act of making your stimuli (your speech) stand out in a way that alerts people's attention.

Some things stand out more than others. In fact, in any given visual scene (like a room), different things will stand out to different people. Taking this further, as people move around a room, different things come into focus for them and stand out to them, while things that had previously stood out fade into their background experience. It seems that there is a fluidity to what stands out in a given scene and that our motion, viewpoint, and time play large parts in making some things stand out more than other things. Idiosyncratic features like personal interests, familiarity, knowledge, and an individual's physical location at a given moment also play strong roles in making things stand out and other things less so. Many factors converge to bring something into the foreground as a salient object.

This is not simply a visual phenomenon, but in fact happens in all of the sense modalities and in multiple domains. Things can stand out to peoples' eyes, ears, noses, mouths, and touch receptors. Things can stand out in time, in space, in awareness, in memory, in routines, in collections, in feelings, and so on. This is a great benefit to artists and experience designers because it means that there are multiple levels on which you can engage your audience—giving them multiple types of salience through a wide variety of formats and media.

But what is this phenomenon at its core? *Why* does anything stand out? Do properties of the things we notice cause them to stand out? Perhaps in part, but the whole picture has greater complexity because it is a relationship between things in the physical world and our abstract inner world. Things stand out because people attend to the stimuli of the world with an attentional system that is fluid and unresting and attention is disproportionately distributed in the world to some things and not others. It is our attention system that causes things to stand out to us, and properties of the things themselves contribute to how we notice them. We'll explore this more as the chapter unfolds.

The study of attention systems has produced a variety of cognitive theories that explain how attention works, most notably Talmy (2000, 2008). But here the focus falls on how attention can be engaged to evoke responses, and how attention works as a tool of cognitive engineering on a rhetorical and semiotic level. One model of attention that helps set up a framework for how to work with other people's attention is Oakley's Greater Attentional System (2009).

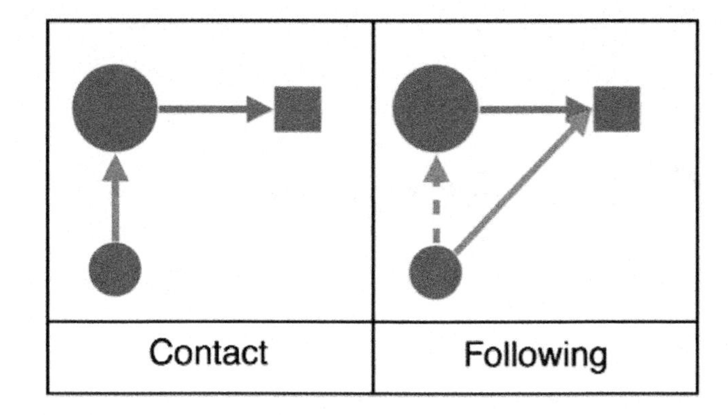

Figure 2. Contact and Following.

This model of attention is broken into three systems: the ***signal* system**, the ***selection* system**, and the ***interpersonal* system**.

The signal system deals with how things stand out, the selection system deals with how we pay attention to things over some length of time, and the interpersonal system deals with the ways that attention works between people and how it might be useful for enabling people to share attention. Let's look at the components of these three systems.

The Signal System

This system consists of two elements, ***alerting*** and ***orienting***, which work together in Oakley's words as "*the sensory and dispositive boundary conditions of human meaning making: they determine that which is significant without being significant in themselves*" (Oakley 2009, 27). Alerting is the active element that helps us notice stimuli and is activated when stimuli present themselves with varying degrees of intensity. When new stimuli present themselves, alerting is the process that helps pick which thing to pay attention to in the scene. Orienting is the element that helps pick stimuli based on spatial, temporal, and cultural criteria. It helps organize stimuli based on value to the situation and on relationships to our acquired frames of reference.

The Selection System

This system consists of three elements, ***detecting***, ***sustaining***, and ***controlling***, which are the elements that help humans manage their own cognition and consciousness (Oakley 2009, 29). Detecting is the element that helps humans see what is relevant to what they are doing, it is a recognition process that relates stimuli to the tasks at hand. Sustaining enables concentration and focus over lengths of time. Controlling allows you to use two processes, *switching* (***within*** the same domain) and *oscillating* (***between*** two domains) to coordinate during an activity.

The Interpersonal System

This system consists of three elements, ***sharing***, ***harmonizing***, ***directing***, which enable people to work together on physical and mental activities. Sharing is the element that permits people to have "*peripheral awareness of another*" (Oakley 2009, 34). Harmonizing is the element that enables joint attention between individuals. Directing is the element that enables one person to manipulate the attention of other people.

Without knowing anything else about the greater attention system, its utility to artists is clear. The very act of creating something designed to evoke a response of some kind depends on being able to capture someone's attention (signal), to control that attention (selection), and to make their attention work for your purposes (interpersonal). This system will be used throughout the remainder of this book as one of the underlying frameworks for building effective experiences.

An Example from Performance Art

Situations can be presented to the audience members that draw on attention patterns from one of the three systems and their elements. For instance, *harmonizing* attention works when individuals are able to engage in joint-attention. Marina Abramovic has explored mutual gaze in several of her performances such as *The Artist Is Present* (MoMA 2010) and the *Eye Gazing Chamber* which was planned for the Marina Abramovic Institute, and has explored synchrony between minds in other projects such as *Measuring the Magic of Mutual Gaze* (Abramovic et al. 2011) and as a rider in 2013 on the *Compatibility Racer* (a project of Silbert et al. 2012). While some of these works depend on technology, others depend on simple face-to-face gazing. This face-to-face harmonization of attention lies at the root of joint-attention (Tomasello 2003; Oakley 2009) and is even the context for the most basic form of language use: the face-to-face conversation (Chafe 1994).

When we apply Oakley's model of attention to Abramovic's performance, we can see that gaze in Abramovic's works is one of the mechanism that *alerts* both parties to the agency of the other and works to *sustain* itself through the performance of mutual gazing as the joint activity between the two parties. Interestingly, this sustaining force pushes aside the *controlling* element (which entails switching and oscillating attention between stimuli) and perpetuates itself by the simple language of a mutual gaze. The mutual gaze bypasses the *sharing* element of the interpersonal system because this is not just a simple case of having peripheral awareness of the other—instead it is the intentional act of *harmonizing* attention that drives this joint activity of mutual gazing.

In the case of the *Compatibility Racer,* a cart is set into motion and its speed is controlled through a brain-computing interface (a pair of EEG headsets and processing software) by two people anticipating and matching the expressions of the other person. This harmonization from mutuality and synchrony works to enable both people to act with vigilance relative to the other person's signals (*alerting, orienting, detecting*) which enables the two parties to *direct* each other's attention in a mutual negotiation of signals toward synchrony of attention and gesture, which then results in the cart being set in motion as the interface finds matches between two people.

Attention is all about relationship: the relationship of the attender to the object of attention, the relationships between signals in the world as they differentiate themselves, and the relationships between people as they coordinate attention. Your work will inherently have some organizational structure that engages human attention in some way, and your work may or may not succeed at directing attention. Make your goal to design the elements of your installation in ways that emphasize a particular system of attention.

Attention is a system and as the attention system operates, the different elements of that system will play their part in the operation, but you can build an experience that isolates one particular element of the system (like *harmonization* in the case of the *Compatibility Racer*). Attention has a working pattern in which different system elements are activated episodically; the alerting element is not always operating, only when something in the environment is acting as a salient stimuli, and then the mind begins to respond to that stimuli and moves on to another element of attention like orienting, or controlling. All of the systems of general attention work together on some level, but many elements are silent in the episodic working of attention. Our ability to focus follows a similar pattern: in order to focus on one thing, by necessity you have to ignore the others.

As you plan for using attention in your experience, determine what functions of attention are necessary for your context and goals. For Abramovic, it seems that switching and oscillating attention (the control element) were detrimental to establishing a mutually sustained gaze, and in terms of the Greater Attention System model, mere sharing of attention would have been a shallow failure of harmonized joint-attention. Removing obstacles to harmonized attention meant removing the options for switching and oscillating. Ensuring deep harmonized attention also meant finding ways to create an environment that made the presence of another person focally explicit rather than merely peripheral (as it is in most social settings). In the intersubjective interaction in *The Artist Is Present,* Abramovic created an environment that made the presence of the other person explicit by presenting patrons with a seat at a table across from her in the center of the gallery. Taking a seat at a table where someone else is sitting renders an explicit awareness of the seated person. It is markedly different from sitting next to a person on the bus where awareness only needs to be peripheral. Instead, walking to the center of the room where there is a single table with two facing chairs and someone sitting in the other chair is explicit and nothing distracts you from being aware of the other person while seated in that chair.

Abramovic removed other obstacles for harmonization by creating a scene in which people would not want to switch or oscillate their attention away from her. Abramovic was thus a magnet for undistracted attention, and several factors commanded this kind of attention:

- She created access to her body, person, and gaze (not normal public behavior).
- She created scarcity by having a show bound by time limits.
- She created sacred space by separating space into waiting and encounter areas.
- She gave value to people by giving them personal attention (acknowledgement).
- She broke social norms to look people in the eye.
- She created a transaction where people spent time in order to receive attention.
- People could not control their attention because the sincerity and personalization of her gaze would not let them switch and oscillate their attention elsewhere. To some people this was uncomfortable while other people found it engaging.

Take the overall goal of your experience and determine which elements of the experience can serve the attention goals that you have. What can be taken away? What must be taken away? What can remain? What must remain? This will begin to help you define the attentional underpinnings of your experience. Next ask yourself if there is any temporality or rhythm of time or sequencing to the flow of attention in the experience. Determine if there is a spatial structure to the flow of attention. Do the temporal flows and the spatial structures have any overlap?

In Abramovic's work, the temporal flow of attention was significantly separated by spatial structure. While waiting in line people did not have eye contact with Marina and so they could oscillate and switch attention, they were onlookers or bystanders (not ratified participants in the mutual gaze). Possibly onlookers switched between looking at Marina and looking at the person in the chair across from her. Possibly they oscillated between attending to the experience in the center of the room and to the length of the line ahead of them or behind them. Or possibly they were attending to their present situation of being in the gallery and then switched their awareness to plans they had for later in the day at work or at home. While standing in the line (which is ***outside*** of the sacred performance space), the patron onlookers saw two participants engaged in direct sustained attention. Importantly, once a patron crossed the boundary into the performance space, their attention became structured by that space for as long as they remained in that space. Oscillation and switching were removed as options by the purity of the experience of receiving direct eye contact and mutual joint-attention. The flow of attention differed vastly depending on spatial location.

While engineering your experience, if you can determine what distracts attention and what heightens attention, then you can chart a flow of attention that unfolds over time and in space. You will create a space that controls and directs attention as the patron moves throughout the space, and their attention becomes part of your toolbox for engineering (more on this in Chapter 4).

At this point, we shift back to notions of salience (or focal elements) and explore how a perceptual structure called **figure-ground organization** manifests itself in our perceptual and conceptual experience.

Objects of Salience: When Things are Focal Points

When something stands out against other things that fade into the background, it is what psychologists call a ***figure***. The background is known as the ***ground***. Figure-ground relationships are probably most familiar to people in the classic Gestalt image of two faces looking at each other where the negative space between them forms a vase. When you are focusing on the vase it is the figure. On the other hand, when your eyes switch to focus on the faces, they become the figure and the vase becomes the background. This oscillation between figure and ground is in fact an oscillation of attention and highlights in a simple way that attention is not a stable phenomenon—it is dynamic and active. Figure-ground organization is not a new idea in 2D art, but thinking about *multi-sensory* figure-ground organization in *durational,* and *path-based works* goes beyond using the concept for composing and critiquing static works. For those of you familiar with figure-ground structure, the new idea here is how this attention pattern occurs during the flow of attention to build up what we think of as experience.

Figure-ground structures enable attention to shift between awareness of the figure and awareness of the background: it's an oscillation of attention. Oscillations of attention are useful points of entry for engaging an audience because they permit a kind of control of audience attention. In order to disrupt attention, alert an audience, and orient their attention, you merely need to create a new figure that disrupts and displaces the old figure. Pickpockets, thieves, illusionists, and con artists have known this for a very long time. Make something new stand out and take the place of what normally stands out in the scene. Think about how this happens in restaurants: all of the diners are sitting at tables facing each other quietly talking over their meals, and then all of a sudden, one of the staff drops a tray of dishes, shattering the glass with a crash. Do all of the diners continue with their conversation? Human tendency causes most people to turn around and look at the scene. This is because the sound of crashing dishes stands out from the relatively calm restaurant scene and it becomes a moment of salience in the attention of the diners. **Salience pulls people away from whatever is sustaining their attention and directs it to the disruption that created that moment of salience.**

To illustrate how banal that concept is, consider that you paid more attention to the last sentence because it was typed in bold and it disrupted the flow of the information by being different from the surrounding non-bolded text. A bold type font, when used in an otherwise non-bolded context blends form and function to create salience. Bold alerts readers to text that has conceptual importance. Bold fonts are used by writers to direct the attention of readers, and it works because it disrupts the pattern of non-bolded text.

This last idea is interesting because it means that we can intentionally create disruptions that capture and direct attention, and so these disruptions act as a kind of communication. Disruptions can work to communicate the intentions of the person who created them. When we use the direction of attention to indicate meaningful things to other people, attention enriches language and becomes a sort of perceptual language. Learning to use this language requires the following: awareness that attention is a system, the ability to appropriately disrupt attention, and the use of disruption that is tied to information that you intend to communicate.

Looking at this idea of using attention to mean something, a framework for *joint-attention* (attention shared between two people onto some third entity) helps to further equip your efforts in using attention in your engineered experiences. Joint-attention is a coordination between two or more people (e.g., you and your audience) and some third entity (object, concept, person, etc.) (Tomasello 2003, 21–28). Our capacity for joint attention emerges in childhood as we learn to understand that someone else intends for us to share attention with them onto some other entity, and we use this capacity throughout our adult life to learn what people mean for us when we don't understand what is going on around us. We use joint-attention to make meaning for and with each other, and it is through this process that we can create situations that enable people to attend to something we mean for them to comprehend or experience. Taking these ideas in line with the notions of joint activities, a more detailed picture emerges on how the skillful coordination of behavior and attention help to build meaning-making elements into the architecture of an engineered experience.

Environments of Non-Salience: Ambient Scenes and Grounds

Backgrounds are just as important as focal points, not just because they enable focal points to stand out, but because they communicate something about the scene whether it is the mood, the context, or a collection of potentially active but not yet salient elements in a complex scene. Some immersive environments bombard the participant with one figure after another figure in rapid succession such that attention is being called upon continually through alerting. Ambient environments are different: they are grounding environments where the focus is less on what creates the ambient environment and more about the feeling that the environment evokes. Think of it like the way that Impressionism tried to recapture the ambient qualities of reality in paintings as a response to the matter-of-factness of photography. In our case, creating ambient environments is kind of like Impressionism for real life experiences instead of paintings.

The immersive sensory texture of an installation qualifies as ambient when the subject of the work are the sensory stimuli themselves acting as contextual surroundings. Ambient in this sense contrasts with object-based work.

Ambient environments can serve a rhetorical function, such as when they act as the background for some other salient commu-

	figure	**ground**
Visual	•movement •generally smaller •simpler in visual field or less complex than background	•stasis •generally larger •complex or made up of many layers in visual field
Auditory	•new sound •contrasting silence	•background noise •background silence •repeating sounds
Tactile	•new texture •discontinuity	•continuity of texture
Olfactory	•new smell •contrasting offensive smell •contrasting pleasant smell	•pervasive smell •background smell •lack of smell
Taste	•first bite •new flavor •pleasant / unpleasant flavor	•successive bites •background flavor •no taste
Temporal	•new event •state change / new state	•habitual event •continuity of state
Spatial	•new environment •new spatial arrangement	•habitual environment •stable spatial arrangement

Table 2. Sensory-Based Figure-Ground Structure.

nication channel, such as performance art, dance, or theater; the environment can tell you something about that specific event and it defines some of the meaning of that event by controlling how you experience that event. When this happens, we think of the ambient environment as establishing the atmosphere or the vibe of an experience, and ambience is an active product of a rhetorical ambient environment.

Ambient environments can also exist on their own without supporting any particular performance at all. When the environment is the work itself, what would normally be background information becomes the primary information. The background is the entire message that is being communicated and so it becomes salient while you actively engage the ambient experience as a figural object. In those cases, the ambient environment is not the background, but the foreground. It is not being used as a ground but as a figure, and this produces moments of contemplation on the environmental conditions presented in the ambient scene.

What Makes a Figure and a Ground?

Talmy (2001) outlined a framework for observing how figure-ground relations in conceptualization and perceptual systems show up in language. Because language is a window into the mind, this model is not limited to language and is useful for visual perception and other sensory-perceptual, spatial, temporal, and cultural modalities. This makes his model particularly useful for the types of engineering activities described in this book.

Talmy argues (2001, 315–16) that primary objects (i.e., figures) are smaller, more mobile, simpler in terms of geometry, more recent in awareness, more important, more relevant, and less immediately perceivable (probably because of their size), but more salient when they finally are perceived. He describes them as more dependent, and claims that their spatial and temporal properties are unknown—by which he means it is an unanchored entity in space and time. He then outlines the characteristics of secondary objects (i.e., grounds) which are near opposites of the figure characteristics, indicating that they are larger, less mobile, more permanent, more complex in terms of geometry, less recent in awareness (presumably as *older* information), less relevant, more immediately perceivable, more independent, and have known spatial and temporal properties that render the ground as a reference to make sense out of the figure's unknown properties—in essence, the ground as the known and obvious entity anchors the figure with its to-be-determined meaningfulness in relationship to the ground.

Figure-ground organization occurs in all types of sensory experience (haptic, visual, auditory, olfactory, gustatory and sensorimotor) (Table 2). It also structures our sense of time and space, the way we feel, the way we speak, and the way we organize information. In all of these types of experiences, the figure is the point of salience and it is contrasted against elements in the background. Calling attention to something is merely a matter of making everything else less interesting so that the element you want to be salient seems like the most interesting thing relative to the other stimuli. In other words, following Talmy, **what you want to stand out becomes salient when you can make it more relevant, more obvious, simpler, and fresher than everything else**. At its heart, figure-ground organization is simply a separation of noise and signal. You could say that the act of something becoming salient is the result of the relationship between the ground and the spatial and temporal properties of the figure becoming known by the perceiver.

The flow of attention is cinematic. Rhythms of information flow and the modulation of attention cause figure-ground organization to constantly shift, such that, as you move through a space, what you notice as figure is constantly disappearing in your line of sight as new figures present themselves in the visual field. Think about driving a car and noticing someone's stupid vanity license plate in front of you, but then you drive past them and notice some roadkill on the side of the road, or a cop hiding behind some bushes, or you see your exit. As you drove past the driver with the stupid vanity license plate, you came to attend to other details along the stretch of road. That stupid vanity license plate which was momentarily a salient figure faded into the background and it was no longer even a ground: it was just gone. But suppose later you notice another vanity license plate and this one stands out to you because it really is pretty clever, even though you don't typically find license plates interesting: as you remember back through all of the stupid vanity license plates that you have seen, this one that you are focusing on because it is pretty clever becomes figural against the ground of all the other stupid plates that are not actively in your focus. We can have figures in our memory that stand out from the ground of the visual field even though they are mental imagery (this becomes useful when designing path-based experiences).

Sensory, Spatial, and Temporal Salience

Our senses and notions of space and time also organize along figure-ground structures. Table 2 suggests some of the distinctions between figures and grounds, but it is not exhaustive (Table 2 sets the groundwork for Tables 3, 4, and 6).

This idea of figures standing out from grounds in sensory and spatial-temporal streams is useful in capturing attention within a work, much like a foreground element in a photograph may stand out from the background and capture the attention of viewers in a particular way. But this static figure-ground relationship needs to be *animated* in order to be meaningful in a *sustained* experience. In the same way that photography evolved into film by sequencing a series of images in succession, dimensional works eventually developed into durational works and installation which involves paths through an exhibit and the linking succession of movements in the exhibit which correlate with photographic stills in a motion picture.

In these path-based and time-based works, viewers experience the exhibit in ways that are similar to the movie camera capturing footage as it moves through the set and scene. Viewers see a *flow* of information, and that flow of information is made up of different scenes, each of which have their own figure-ground organization. As the collection of scenes is encountered through the camera's movement, or through the viewer's movement through the scenes, many salient figures enter the scene in succession and they stand out from the background, even as the background changes as a result of the camera or the viewer moving into a new environment. It is this viewer-centric approach to attention that makes figure-ground relationships so important to understand. Viewers move through installations at their own paces, with their own goals, motivations, and interest levels. It is vital to think of viewer experience as cinematic where the viewer is the camera and the experience is the film.

There is always variation in viewer attention spans as well as variation in bodily structure. Everyone is a different height, some people are exhausted from a busy day on their feet, some people have to go to the bathroom and so they're distracted or rushed, and so on. Otherwise observant people might be more interested in the date they brought to the museum, or maybe the gallery happens to be the backdrop for some business entertaining potential clients, and nobody really cares about art at all.

You can't write a book that answers all of the critiques that people will throw at it, and in the same way, you can't make an installation that appeals to everyone that experiences it. That means you need to understand how to layer attention in the exhibit for multiple types of people. Just like a good non-fiction book is accessible to novices and experts, aim to layer your installation to appeal to a continuum of types of potential experiencers: accidental experiencers, casual experiencers, intentional experiencers, disinterested experiencers, and unaware experiencers.

Let's get back to the idea of the flow of attention—specifically, how figures and grounds *sequence* in events and extended periods of time. Things stand out in time, and a rhythmic nature of figure-ground relationships can be understood as a way to organize temporal experience into a flow.

Figure-ground patterns happen in our experience of time. For example, when you go on vacation to Paris and stay in a hotel, the hotel becomes a sort of base-camp for your exploration of the city. You wake up and leave the hotel and visit *la tour Eiffel.* You come back to the hotel after your visit for a moment and perhaps go out for dinner soon after only to return back to the hotel for the night. In the morning, you leave the hotel and walk down the *Champs-Élysées* to see the *Arc de Triomphe,* to visit the *Louvre,* then you go back to the hotel. The next day you walk along the *Seine* and take a class in French cooking at a cooking school, then you go back to the hotel. The next day you walk around shopping in boutiques and then you go back to the hotel. Each day your experience of Paris begins and ends with your experience of the hotel and this acts to stabilize the trip as you experience the "highlights" of the trip. The hotel is the stable constant *ground* that helps to make the day trips into memorable *figures*. There is a frequency to the flow of the trip and the hotel is the baseline, and each separate event is a peak that spikes away from the baseline before returning to the baseline again, much like the peaks of a waveform readout of a heart rate monitor—the exciting things make the peaks, and the everyday steady state is the baseline.

Over the course of a Paris vacation experience, this back and forth temporal ground increases task-oriented salience on the elements that are different (figures) and it decreases attentional structure on the hotel (ground). This effect has been observed in tourist descriptions of an experience, identifying differences in simulated motion based on whether the descriptions are of habitual (hotel) or institutional (tourist sites) settings (Dewey 2012). Without stretching this too far, it is possible that this kind of figure-ground oscillation in our daily schedule affects how we think about spaces and affects how we describe our experiences. In other words, how we understand an experience may be linked to the figure-ground relationships that exist in that experience. Going on that assumption, it seems reasonable that perhaps these time-based figure-ground relationships in long-durational events (like vacations) also structure shorter events (like a gallery visit).

In the engineered experience, providing a stable element throughout the experience can act as a ground that helps people increase task-oriented attention on figures. Figures that are created using time-based organization can build a rhythm, which might be used for multiple purposes such that, in your designed experience, a given location in time can also provide an emotional trigger, solicit audience engagement, or provide some sensory data that characterizes that moment in time with a sensory anchor, such as a particular smell or a color.

Unless you are trying to build a background ambient feeling, you want to put the meaningful elements of an experience in the temporal and spatial figures because that is where they will be noticed.

If figure-ground organization occurs in how people make sense out of experiences, and if it can be seen to stem from where people spend their time, then in designing experiences, time-based figures should be moments that are paired with the meaningful elements of the experience (like triggers, decision points, knowledge transfer, narrative advancing devices, memory anchors, etc.). Time-based grounds should be the stable contextual structure of the experience providing the background and contrast to the figures, establishing more salience on the figures by blending into the background. In an engineered experience the time-based ground accounts for what people do most of the time during the experience. The time-based figures point out what people only do infrequently during their experience. Make the most of temporal figures by designing them to be moments of intensity in the experience because their intensity will cause them to stand out.

The Rule of Thirds: Our Common Ground

Some things stand out because of their inherent properties and contexts, some things stand out because people want them to stand out or design them to stand out, but there is a cultural element that also causes things to stand out. We also attend to things because they are part of culturally-learned visual aesthetic preferences. Consider the *Rule of Thirds* where a visual scene (e.g., a photograph) is divided into a grid with four intersecting lines and the preferred location for a subject that you intend to stand out is wherever two of the lines cross. These intersections turn a subject into a point of salience, and that salience is a separation of a figure from a broader ground.

The Rule of Thirds is a critical element of sensory composition as you design and hack into experience. And if you think about experience design like a film (which is basically a sequence of still images) the rule of thirds makes itself useful as you design the individual moments or snapshots of the experience.

We're using the Rule of Thirds in this book as one of the connection points that bridge between art and science. It's a tool that perceptual science and artists both make use of in their different approaches to studying and exploring perception. The Rule of Thirds is a culturally derived method of composition, there is nothing inherently "correct" about it. Some people argue that perception naturally organizes according to the rule of thirds because it is something we are born with. This is enticing, but other people argue that instead of being born with this ability, we've evolved culturally to prefer the Rule of Thirds in visual works. Whether or not the rule of thirds is innate doesn't change the fact that we use the rule of thirds now and it helps us to see (without thinking about it) what the artist or photographer wanted us to see as salient. In this way, the Rule of Thirds is a rhetorical device to make meaning, kind of like a label that says, "this is what I want you to experience as important," and you see it and it makes sense to you.

Western culture has used the Rule of Thirds to make certain things stand out in visual structure. This is an information-structuring device that Western audiences are particularly aware of in the way they consume visual information. Whether or not the Rule of Thirds is universal is still out for debate. However, a universal audience encountering your use of the rule of thirds as the structural armature of the experience you design will certainly evoke a variety of perspectives on your designed experience, and that is good because it makes your experience interesting and engaging.

If you are not familiar with the Rule of Thirds, perhaps you are familiar with the idea of a grid (Figure 3) being laid over an image. When you lay a grid over an image, the salience is often located a third of the way from the left and a third of the way from the bottom of the image. This is a way of establishing a kind of visual harmony in the image. But salience can happen in other places in the grid, as it works in any of the intersections marked with an x in the grid (Figure 4).

You can apply the Rule of Thirds to more of the world than just visual imagery. The Rule of Thirds provides a level of common ground that we can use to hack into experience by placing interventions in sequences that follow the rhythm of the Rule of Thirds in non-visual space. The Rule of Thirds is a spatial rhythm and provides a kind of timing for the sequencing of information in many domains. Because it is a sequence, if it is a sequence of the right kind of information (i.e., presented using figure-ground organization), then the Rule of Thirds can apply equally well in the other sensory domains (sound, taste, touch, smell) as well as space, time, and even narrative. In fact, it is a way to point out figures and grounds in the sensory composition of a designed experience.

Since this is a rhythm, it might be easiest to demonstrate the applicability of the Rule of Thirds to music which is a blend of sound and time. Divide a given piece of music into four segments of time. The third segment of time will probably feature some kind of discernible musical change (e.g., possibly a key change, possibly a shift in tempo, possibly a lyrical shift or a new voice entering the song). Some music does this because that's how it has always been

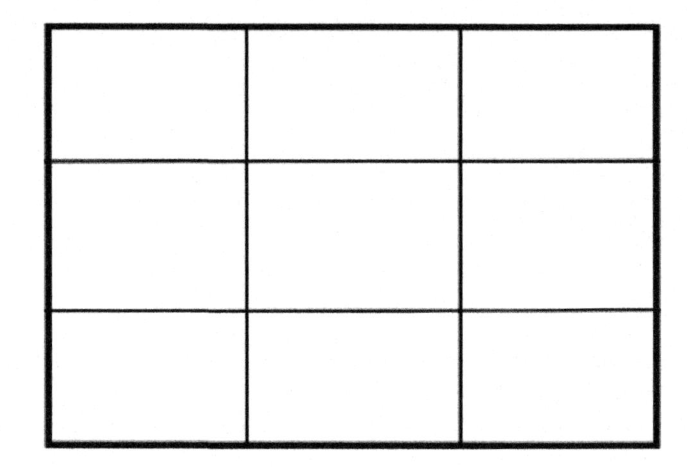

Figure 3. Rule of Thirds Grid.

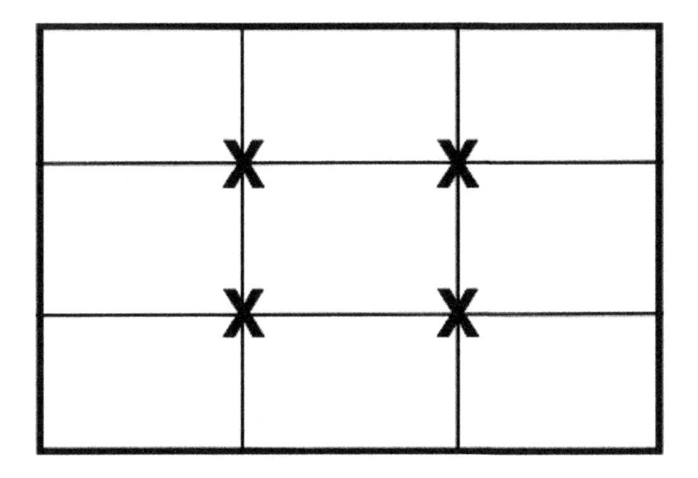

done. Other music does this because it blends form and function to leveraging a moment in which the listener is expecting to hear one musical statement, but then is surprised by hearing a new statement. Even the most basic song like Happy Birthday does this with a key change, tempo change, and change in inflection during the third phrase, "Happy Birthday, dear Jacques." So the Rule of Thirds is one way to build up anticipation. The music means something for the listener at that moment. "Hey, pay attention, we're climaxing here."

At the beginning of certain types of classical music, the musical statement stands out from silence. The first quarter of a piece of music is all new. But then some of the themes repeat themselves in the second quarter and the theme begins to fade into the background. Once the theme is in the background, something new can come onto the scene and stand out in contrast against that background as a salient figure in the third quarter (following the Rule of Thirds). Often the music returns to repeat the original theme in the fourth quarter. In pop music the Rule of Thirds appears this way: the song begins with the first verse in the first quarter, then moves into the chorus in the second quarter, and repeats. Then the unexpected break in the pattern occurs with the bridge, before moving into the fourth quarter, which repeats verses and/or the chorus. David Bowie's song "Golden Years" does this particularly well, especially considering the repetition of the second chorus as the fourth quarter, just after the bridge.

Consider literature. Grab a fiction book and find the pages that are in the third quarter of the book. In traditional narrative structures, very likely this third quarter coincides with the climax of the narrative arc. In some sense, this can be seen in the *Bible* when Jesus Christ enters the scene in the third act (or three quarters of the way through the book). (For more on this pacing of information, see Chapter 5.)

You can apply this same Rule of Thirds sequencing in the other sensory domains to make certain things stand out in the designed experience. You will hack into participant experience by subtly borrowing a familiar frame of salience from visual culture and adapting it into some other sensory system.

Figure 4. Salience in the Rule of Thirds.

3

Experience Hacking Is Fundamentally a Form of Design

We've started to see how to break up experience into some basic elements so that we can work with those elements as doors for hacking. We've started to see how basic patterns of human attention can be used to drive an experience and to pull the experiencer along. Think of your content and medium as the clay that is placed upon an armature during sculpture: these are the ideas and forms that the viewer obviously interacts with, and they are shaped with tools from your cognitive toolbox (more on this in Section 2). But underneath that clay is an armature, a supporting structure that is the skeleton of the sculpture. It is the same with designed experiences where the organizing concept, sequence of events, and the spatial layout act as the armature of the experience. In Section 2 you'll learn more about the tools in the cognitive toolbox, but at this point, before going any deeper into the elements of experience and the use of attention in an experience, we need to better understand how to build a strong armature for that experience. Like Chapters 1 and 2, this introductory chapter on armature begins an exploration that will continue throughout the more technical chapters of this book (starting at Chapter 4).

If attention drives much of our selected experience, then it is useful to understand how attention plays out during experience. Figure 1 laid out a big-picture model of hacking experience. In terms of aesthetics and composition, we usually encounter ideas like figure-ground organization and the rule of thirds in discussions of 2D art forms, but as we've seen in Chapter 2, we can activate these tools in experience design to structure events with sequences and oscillations of salience to capture and direct attention. We'll see in chapter 4 how those sequences and oscillations can incorporate viewpoint, sensory stimuli, use of time, and the use of emotion. All of this occurs with the underlying structure of narrative, and it will always be kind of an experiment, as well as an attempt to evoke responses. Spatially all of these interventions with the senses and emotions and attention will happen in some sort of path form. Paths can be divided into segments and nodes and this spatial structure will coordinate with the already mentioned narrative arc that serves to tie the story of the experience together. Not only does narrative help the participant in the moment of experiencing your design, but it presents the design in a format that they can use to build a memory in story form. Your use of story coupled with space will anchor that experience for many people.

By designing an experience you engage in what Merlin Donald calls ***cognitive engineering*** (2009). Regardless of whether your purpose is creating art or cooking a meal, it is fundamentally a type of design. We're approaching hacking as design, whether it is an installation, participatory/interactive environments, performance art, curatorial practice, architecture, landscape architecture, complex durational works, or works requiring new models of documentation, we should learn some lessons from disciplines like engineering or design research and borrow the frameworks of engineering where they are useful to our purposes.

Engineering works to solve the form/function debate as a basic problem. Every designed solution has a corresponding function that makes it a solution, and that function emerges out of the problem that the solution solves. In design we think about the elements that are *necessary* to solve a problem. Think of design as an approach to structure something in a way that satisfies the requirements of the functional goal you are designing for.

When the task of the artist is to create experiences that evoke responses, those experiences (and the process and research involved in creating those experiences) can be looked at as a kind of design problem. Creating experiences, whether those experiences are installations or paths through a garden, have a number of commonalities with what Rittel & Webber (1973) described as "wicked problems." Wicked problems in design are problems that are indeterminate, slippery, and multifaceted. They are the kinds of problems where one solution brings other problems with it. Where each problem is actually the symptom of another problem, where nobody knows what the range of solutions can be, and no one regulates designed solutions.

Solutions to wicked problems can't be evaluated by categories of truth and falsehood, but are only appropriately judged on their ability to be satisfactory and where they fall on a continuum of being good or bad at satisfying the problem. The task of the artist in designing an experience is much like these types of problems. The solution you implement isn't true or false, but it can be evaluated on how well it satisfies your goal. Certain decisions you make will uncover new problems, and you won't be able to address all problems. The only rule for stopping in the design process is when you feel that you have satisfied the design goals, otherwise the problem itself doesn't provide any "stopping rules."

Wicked problems aren't just problems, they're also an opportunity for the artist. When an artist defines a problem, or problematizes something in the world and begins to address it as subject and content in their practice, it is up to the artist to define the list of possible solutions to that problem. This is a great place for speculative design to take a foothold in your practice. Besides being a form of "social dreaming" (Dunne and Raby 2013), speculative design is really an attempt to solve some wicked problem, or at least to start a conversation about that problem, and even that conversation about the problem is realized as a wicked problem too.

Perhaps the most useful framing of a wicked problem is Rittel and Webber's statement "*there is no definitive formulation of a wicked problem*." They continue,

> For any given tame problem, an exhaustive formulation can be stated containing all the information the problem-solver needs for understanding and solving the problem....This is not possible with wicked-problems. The information needed to understand the problem depends upon one's idea for solving it...in order to describe a wicked-problem in sufficient detail, one has to develop an exhaustive inventory of all conceivable solutions ahead of time. (Rittel & Webber 1973, 161)

This kind of information collecting strengthens your personal approach to experience hacking by giving you a more complete picture of the problem itself and also illuminates a range of possible solutions to that problem. You can collect information as a way to think about the design process, allowing you as the designer to reflect on the process of experience design as you collect all of the information related to that design. But this is getting ahead of ourselves. Let's relate this back to experience design by looking back to Chapter 1 where Spradley's nine dimensions of experience were introduced (dimensions such as *space, object, act, activity, event, time, actor, goal,* and *feeling*). One way of going about designing a complex experience is to go domain-by-domain and make lists of the types of components your experience will have or needs to have. Then go back over that list and identify the different relationships that exist between those components. By acting like an anthropologist and listing out the different elements in your design, you'll be gathering together a set of facts that you can use to propose multiple design solutions. You can think about which solutions will work best and then you can begin to turn that solution into a story. And that's what experience design is about, **telling a story through the experience**. Once you've listed out the elements of your proposed design, you'll begin to see how everything in your design comes together, and it will give you an idea of where to start telling that story.

Why Bother with All of This, Anyway?

If this discussion about solution design seems heavy and pedantic, you don't have to use this method. But for some complex experiences, using a method like this will be helpful, if not in the planning process, then as a tool to make documentation about your project much easier. Documenting a complex work like an installation or a durational experience takes time and diligence on your part and on the part of your curator or gallerist, and being able to offer them your planning notes (in an edited form) will give them something to work from. It's always easier for someone to tell your story if you tell it to them first, and sometimes the best way to know your own story is to do a lot of leg work of converting ideas in your head into concrete sentences about those ideas. Also, don't kid yourself: it helps to have a systematic approach to how you put your works together too, because it builds cohesion and transforms your individual works into a body of work.

A model from systems engineering called **requirements management** provides another tool for identifying the things that have to be in place in order for a design to satisfy its functional goal. Identifying the requirements for your design to be successful ultimately depends upon your ability to clearly state the goal of the design and to check to see how each component of the design does or does not contribute to reaching that goal. What follows here is an outline of a possible requirements analysis.

Requirements for Design Thinking

Lots of art, and all design begins with a question. Sometimes the question is a riddle that needs to be answered. Other times it is a problem that needs a solution. Still other times the question is how to get people to do certain things. All of your work stems from your initial question, and even if you follow tangents and shift course midway through your project, the initial question is the starting point. It helps to be able to state your initial question or idea in a simple way that enables you to further explore that question. No matter where your question comes from, make sure that you can state it in a clear way.

Let's say that you have a sense of the massiveness of all the time that has already passed since the Big Bang and you want to convey the thrust of that time in such a way that people can feel the same or similar emotional weight of that deep temporality.

Maybe your question is something large like: **what is time?**

That is a big question. Start building more specific questions about that question. Personalize it and explore your relationship to the question:

- What does time mean to me?
- How do I experience time?
- What does my body feel like when I think about time?
- Do certain situations make me think about time differently?
- What triggers my experience of the depth of time?
- Do any specific places cause me to feel more connected to deep time?
- Why do those places trigger these feelings?

Questions like these will help you get the sense of what you feel about your big question and they might help you come to a point where you can articulate why you want to address your big question in your art. These questions become vital if you want to generalize your experience and make it accessible to your audience, because these questions form the basis for beginning the process of hacking experience, especially as your experience is designed to provide the answers to these questions. People discover the answers to these questions as the designed experience evokes responses for individuals, whether emotional, visceral, reflective, or cognitive.

Now that you've fleshed out some of the ways you feel about the big question, succinctly restate your original question again, this time informed by your supplemental relationship questions. Turn your big question into some solvable problem and say something about how you think it can be solved. By problematizing the question, you can start to reverse-engineer solutions and see whether or not a particular solution provides hackable doorways into the cognitive structures of experience. This act of reverse-engineering goes smoother if you use a simplified form of a process that engineers call *requirements management.*

Requirements management looks at all of the elements that go into the design and engineering of some situation, product, service, or environment. It's a process that specifies exactly what needs to happen, when, where, why, and how. If you borrow from this method in the way you produce art, you have access to a tool for organizing your practice that, as we'll see in Chapter 7, also provides a robust form of documentation. This documentation as an artifact of the exhibit is like a field guide or a playbook for managing the complexity of your work and will be useful both for you in the production of your work and for everyone else involved: studio assistants, manufacturers, gallerists, curators, grant committees, the viewing public, your portfolio, and eventually the archivists.

For something to be a "requirement," it can't be optional—that's why they are called *requirements*. They are the elements that *must* be in place in order for something to function according to plan. They provide the roadmap for success in the experience engineering process. You don't need to design a complex experience to benefit from requirements. Anything you design benefits from asking questions about what the design requires. **This is because framing a problem in terms of what it takes to answer that problem helps solve the problem.** Beyond solving the problem, framing a problem by specifying the requirement to solving that problem defines the solution so clearly that it makes placement easier because clarity helps people make sense out of complexity. Requirements establish the test to prove whether or not you've met your goal in solving some particular problem. Let's move along and start the process of identifying requirements.

Identifying Requirements

In order to begin determining requirements you have to know what situation you are trying to address with your engineered experience. This is why having a clear statement of the question you are addressing is helpful. State the situation you are addressing in a concise but complete sentence. For example, if you want to help people understand deep time, start off with a sentence describing the situation you want to address, like this:

> *People have a hard time comprehending geologic deep time since they don't experience time periods that are long enough to help them comprehend long time periods like thousands, millions, and billions of years. In order to help people comprehend deep time they have to have some way to relate to time periods outside of their normal life experience.*

Next, look at the sentence and list which parts are the more challenging parts and which are the less challenging parts. Once you have them separated, start with the easier list and try to figure out what *must* be part of the design process for the experience to have value for the people in the experience. Which functions and elements have to come together in an experience to solve the problem in your situation? Make a note of which functions and elements *must* be present to have a meaningful experience. Move on to the harder list and break down the challenging parts into lists of functions and elements that *must* be in place to solve the problem in a meaningful way. Look at both lists side-by-side.

Are there things on your lists that you can take away and still create an experience that successfully addresses the problems in the situation sentence? Does everything on the list have a critical function in creating the experience? If the design can be implemented without one of the items on your list, then it is not a design requirement, so take it off the list. "Requirement" strongly implies that something cannot be done without some required element. Design requirements are the things that *must* be part of the design for the experience to successfully fulfill the design. They are the Occam's Razor of experience engineering. Before you begin the requirements process, know that requirements never specify a mechanism for how they should be accomplished—they only highlight the problem, not the solution. Any number of solutions may work for a given requirement, and that will be up to you to figure out as you move along the design process.

State the requirements in simple sentences, and create a new sentence for each requirement. Express the force of the requirement with strong modal verbs like must, shall, will, to remind you that they are obligatory and not optional in the design process. For example:

— Long time periods must be presented in a way that makes sense to participants.
— Participants must be able to comprehend deep time.
— Participants must be able to think of time outside of their normal life experience.

Avoid the weaker modal verbs (*can, might, should*) because they suggest less necessity. Remember, you only want essentials in this list.

Go back to the original descriptive paragraph and list each aspect with its own line. Use a separate requirement statement for each aspect of the problem you are solving. For example, people have difficult times relating to thousands, millions, and billions of years. Designing an experience that helps people relate to those different time scales will be easier if you have to find ways to separate out the different scales. For example:

— Participants must relate to thousand year time scales.
— Participants must relate to million year time scales.
— Participants must relate to billion year time scales.

This is better than lumping the different goals together in a single sentence, such as:

— Participants must relate to thousand year, million year, and billion year time scales.

This commitment to clarity will help you track whether or not your goals are being met and will help you identify influencing factors in your finished product.

Make each of these requirements traceable throughout your process of designing the experience; that way you will be able to see whether or not the solutions you design actually tie back to the original problem. Most importantly, find a way to confirm and verify that a requirement is met in the final solution. If a requirement is stated properly this will be easy to determine. For example, if a person cannot relate to thousand year time scales, then obviously the requirement was not met. Test whether a person is able to relate to a thousand-year time scale by incorporating a challenge inside of the experience. If the visitor successfully accomplishes the challenge then you will know if your requirement is met in the experience. The final test for a requirement is if you can turn it into a yes or no question and the answer is "yes":

— Are long periods of time presented to participants in a way that makes sense to them?

Semken et al. (2009) set out to help people understand the geologic time scales that are seen in the geologic formations of the Grand Canyon by making them relatable to people in their physical experience of space. They designed a project to create a timeline called *The Trail of Time* at the Grand Canyon National Park which enables park visitors to adjust their personal understanding of human-scale time (years and decades) to geologic-scale time (millions of years). Visitors experience this by walking along a trail that is marked out like a linear timeline with intervals which correspond to compressions of time as relative to physical distances. This means that one meter on the trail can be used to designate a period of time and thus space can be used metaphorically to represent time. One key element of their design recruits meaningful time periods from a person's life to help them visualize time in spatial form. They chose birthdays as meaningful time periods because everyone has a birthday. As people began their walk along the Trail of Time, they first find their birthdate along the timeline. By giving people a sense for how much time maps to a particular distance, the designers were able to use meaningful spatial distance to help people make sense out of non-meaningful distances (e.g., how someone's age corresponded to the length of a million years on the timeline).

They asked participants questions about their experience relative to their position on the time line and participants were able to track their temporal location in deep time by thinking about their spatial location on the physical time line. Because participants were able to locate themselves in deep time (which was not a kind of time they otherwise had exposure to in real life) by reasoning with physical space on the time line (which they could relate to) the project design requirements were met and participants were able to think about abstract concepts in terms of concrete experiences.

If the designers identify design requirements, such as "*Long time periods must be presented in a way that makes sense to participants,*" the task of solving the problem contained in the requirement is both simplified and testable. If participants cannot make sense out of long time periods, then the solution fails the test and the designers would have to begin again. It seems that the designers found a successful solution by enabling the participants to relate a known length of time (the distance between now and their first birthday on the timeline) to an unknown length of time (the distance corresponding to a million years on the timeline).

The Bottom Line

If you clearly state the problem or situation you are trying to solve/build with your experience, it is easier to recognize the required elements that must be in place in order for your experience to successfully solve the problem/create the desired environment. If you can specify what those required elements are, you can design features that satisfy those requirements and you can build in methods to verify that the requirements have been met. Being able to verify that requirements have been met provides you with a measurement of how successful your project/experience was. This information does two things: 1) it helps you learn how to design better the next time, and 2) it provides you with quantitative evidence for the value of your experience because you are able to measure the response that you evoked with your experience. Both of these results have monetary consequences. As you get better, it saves you time and you know where you get the most value for your effort. Also, as you are able to quantify results, you know the cost of each potential solution which when writing budgets and proposing projects helps demonstrate that you know what to do with grant money. This in turn helps build a track record for your reputation as a good steward of financial resources which leads to future trust and more funded opportunities.

Experience Spaces

Another armature for building an experience is the physical space in which the experience unfolds. We can start with three basic spatial arrangements: object-based, path-based, and landscape-based (like environments & atmospheres).

Let's look at each of these in more depth.

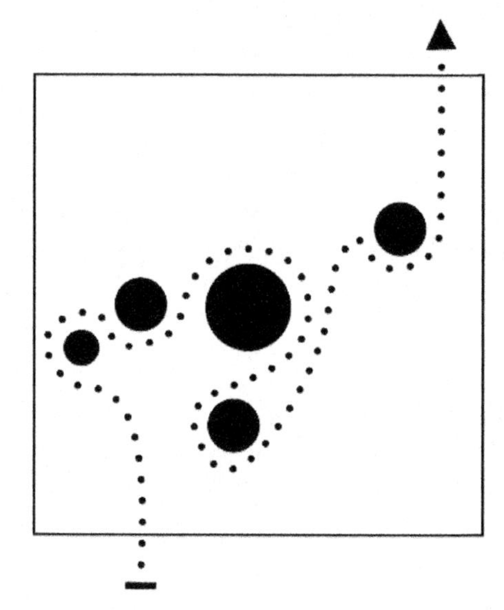

Figure 5. Object-Based Environment.

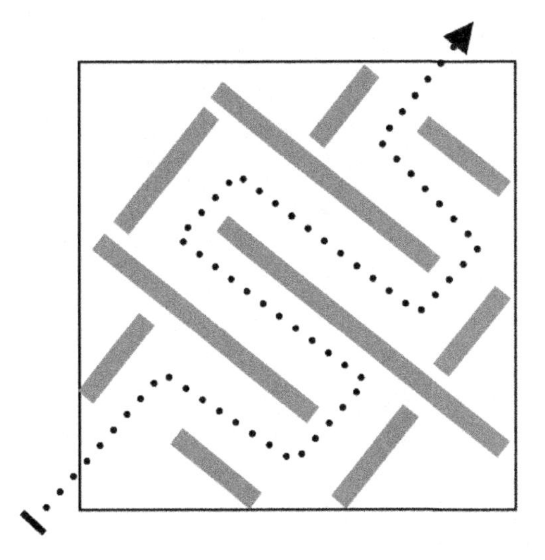

Object-Based

The object-based environment (Figure 5) consists of a sculptural object, station, booth, or a defined space within a larger environment. Object-based templates provide a direct spectator viewpoint onto some object (whether this is visual or some other sensory spectatorship).

The spectator nature of the object-based environment can evolve into participant viewpoints as the participants go deeper and deeper into concentration on the object. For example, a headphone-equipped listening station (Station A) in a gallery setting may involve the listener hearing sounds from another listening station in the gallery (Station B) through the headphones (spectatorship), and underneath of that top channel of station B sounds, a second subtle channel is mixed into the stream of sound that comes from a microphone within station A. During the pauses in sound from Station B, the listener in Station A begins to realize that they are hearing sounds that they are producing (a kind of participation).

Interventions designed using the object-based template can occupy the center of a room, be scattered around a room, or be wall mounted. This is a familiar type of template for galleries and museums.

Path-Based

Think of path-based environments (Figure 6) with the phrase "the journey is the reward" where the journey correlates to the path shape (this stands out from the object-based space in which the journey connects the rewards). The path shape and the elements that define the path shape are the valuable parts of the path-based experience. Something about the path itself will be the content and subject that you communicate to your audience.

Path-based environments provide a linear element to the environment whether it is a path, a route, a maze, a labyrinth, or even a walk protocol that isn't tied to a specific place. Path-based templates include the choice path and the prescribed path. They provide a participant viewpoint that may or may not include moments of spectatorship along the way. All human experiences of space (including the object-based and environmental based spaces shown here) involve some sort of path, but the path in the path-based design is an armature for the experience, it both supports and delivers the content of the experience.

Choice paths allow participants to make their own path through the environment in the order they choose and at the pace they decide. With the exception of mazes and labyrinths, choice paths are unconstrained by architectural and structural elements. However, compositional strategies can be used in the design of the environment to direct attention to try to influence path choice. This would work well in settings where you would like to track audience decisions as they move through the experience.

Prescribed paths work well with sequential organizations (whether or not they are chronological), and participants move through the environment according to a pre-determined path. You might use these types of paths to give participants information along the path, such as details that help tell the story. Paths that are prescribed generally follow a progression of some sort, like

Figure 6. Path-Based Environment.

the progression of a narrative, or possibly a chronological order of events, or the evolution of an idea from inception to execution. (see Chapter 5 for information about how to align narrative with spatial zones along a path)

Landscape-Based: Environments & Atmospheres

Landscape-based environments provide some repetition of pattern throughout a space and can provide zones for engagement whether through ambient environmental conditions, soundscapes, smellscapes, emotional zones, remoteness, or even landscapes in the traditional geographic sense. Landscape-based templates can use both spectator and participant viewpoint in a designed or non-designed manner. A path through a landscape can give, take away, or restrict viewpoints (see Tool #3 for a viewpoint discussion about remoteness), or it can be left up to the participant entirely to change which viewpoints they take.

Quadrophonic installations (Figure 7) are a good example of the landscape-based atmospheric template. In this type of installation, four channels of sound come from four different speakers set in a cross pattern and the zone of sound coming out of each of the speakers mixes together in the center of a sonic Venn diagram of sorts. As a participant walks into the different zones of sound, their position determines which mix of sound they hear from the combined channels. Walking toward the center they hear each channel equally, but as they move toward one speaker or even between speakers, the sounds they hear shift toward the channels in which they are immersed. Sound can be used to direct attention and lead people's behavior in this type of installation if during the experience a certain channel of sound stands out from the other three channels. It can attract attention toward the speaker and encourage the listener to move into or away from that channel zone.

Another possibility for this type of installation is to string together a number of quadrophonic installations in a large enough space and provide a prescribed path through those installations to build a sequence of sound that tells a story. (See Chapter 5 for information about how to align narrative with spatial zones along a path.)

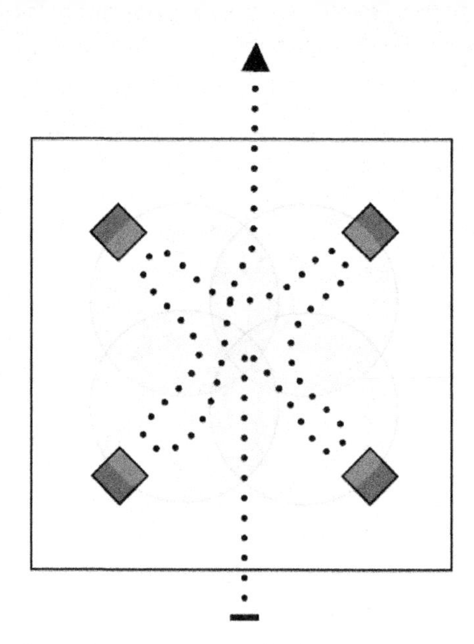

In the landscape-based environment, the *between* spaces (e.g., liminal, interstitial) and peripheral spaces (marginal) provide additional locations for use in meaningful ways. Don't overlook these spaces whether you want to use them to explore the nature of what between spaces and peripheral spaces have to offer environments and sensory ecologies, or whether they become locations for extra, but hidden, content that enriches some element of your narrative or experiment design. When the landscape-based environment is primarily a visual environment, this between space and peripheral space becomes a good location for views and vistas out onto something (spectatorship), creating a kind of porosity in the environment. Also, think of the *between* spaces as cracks that can be used as portals to view something otherwise hidden from sight. For instance, the gaps between the movable walls in Figure 6 allow visitors to catch a glimpse of what is behind the walls. By maximizing these spaces as portals that provide views onto or into something that gives new information about the path, you can build orientation and disorientation in the visitor as a rhetorical effect (see more on this in Tool #3). If you are using narrative design, can you use these *between* spaces as a narrative device to convey information or

Figure 7. Landscape-Based Environments and Atmospheres.

advance the plot forward? (see more on narrative design in Chapter 5).

Think of these gaps and between spaces as the porosity of the space, what can move through these gaps? Information and stimuli can come *into* the path space through these gaps, but it is not unidirectional, visitors can respond and take action through the gaps on objects and intervention elements located outside of the path space. These gaps are bridges between the inside and the outside of the path space and they have their own ecologies of action and energy flow.

The Rule of Thirds in Assembling the Spatial Structure of Multisensory Environments

Think of any space from a spectator viewpoint as if it were a bird's eye view. Now think about that view of the experience space as photographic image. Divide it into four sections and following the Rule of Thirds to locate where the points of visual salience are located (in the third quadrant in any direction). Trace over this image the path trajectory of people moving through the space during their experience. At the points where their trajectory crosses paths with the point of salience from the bird's eye view, create an experientially salient moment. Create a sound installation, or an olfactory stimulation, or use shadows and lights in ways to blind and dazzle or illuminate new vistas. The point is to adapt the rule of thirds to other sense modalities: sonic, haptic, gustatory, olfactory—you can modulate one layer of salience in spatial settings by looking at the plan view of a space and then creating points of salience that match with the Rule of Thirds spatially.

Paths themselves have conceptual points of salience that may or may not match up with the structural-spatial point of salience from the bird's eye view of the environment. It may not be possible to make the most meaningful element of the path match up with the Rule of Thirds, but you can at least try to coordinate content and spatial organization to build in a rhetorical structure that matches the spatial flow of attention. Start by determining the part of the installation that you want to feel like the climax and then see if you can align it with some sort of path that puts it three quarters of the way through the installation. You might not be able to lay a rule of thirds grid onto the floor plan of your experience and match up the climax with the salient points in the rule of thirds grid, but it should be possible to put the climax three quarters of the way into the experience, unless for rhetorical reasons you want the climax to come earlier in the experience of the installation.

In a poetic sense, the cinematic experience of life is basically your motion along an assemblage of paths (like streets, sidewalks, paths through the room) that strings together nodes of events like beads on a necklace. Paths tie together the unrelated events that you experience in life. Paths structure your life in ways that nothing else can because paths provide and take away different viewpoints. As you go about your day-to-day activities, the path gives you perspectives on the world and that perspective helps you make decisions about the world. When something in your environment changes, it shapes the way that you think about that environment. Paths are one tool that help shape the environment. Think about sculpture in the round, your path around that object opens up new perspectives on the sculptural object, what you can see of that object changes and it shapes how you think about that object. The same is true on larger scales stretched out over time: the longer you stay in Paris the more you know about Paris because you've spent more time moving through Paris and watching it unfold as you put one foot in front of the other. You'll know more about Paris both in breadth as you come to see more new areas of Paris, and in depth as you spend more time in familiar areas.

Do not underestimate the importance of paths as a structuring tool, as they can become a rhetorical tool for designing experiences that tell a story, shape a perspective, help build a memory, and even evoke emotional effects.

Throughout the remainder of this book, we'll look at paths as a structural armature and see how they tie together the different elements of your installation. At this point we'll move on to look at some of the tools you can use to design the elements of your installation and we'll return to paths intermittently along the way and then cover path design in full in Chapter 6.

4

Your Cognitive Toolbox

This section introduces a set of tools for shaping the way visitors conceptualize the designed experience.

Our conceptualization of an experience is tightly integrated with spatial and temporal organization. Modifying spatial and temporal elements will shape the way people conceptualize an experience, and the following tools relate directly to the cognitive processes at play as people move through time and space.

Tools like image schemas, cognitive simulation, mental imagery, viewpoint, embodiment, motion, and perception converge and integrate in our reasoning about time and space. These tools also integrate in our emotions and they structure the language we use to describe an experience.

These tools will help you structure the way you use attention in your intervention and will provide a way to blend together the spatial structure of your intervention with the content that you want to convey. Each of these concepts have their own dedicated descriptions, but they also each feed into each other. It makes sense to slowly incorporate these concepts into the text where they feed into each other. You'll read about these tools before you actually come to their respective dedicated descriptions, this approach helps you slowly build up familiarity with concepts, and then the dedicated descriptions provide the answers to the questions that the gradual introduction of concepts raised along the way. So, use this overview as a glossary if you get confused by a term before you reach its complete description. With that in mind, let's briefly look at these concepts before exploring how to use them:

Language is a cognitive interface that can help us see something about the way that a person processes an experience or piece of information. We can also use language to shape the way a person experiences the designed world. Language is one of the easiest ways that we can describe our experience and it is used for structuring our experiences through the use of embodied concepts. Language is particularly useful in translating abstract experience into concrete embodied concepts through the use of schematic structures and conceptual metaphors. Language also makes a good interface for penetrating the mind during the hacking activity by way of prompts and didactic scripts that frame the experienced scene. All of the following tools show up in the structure of our physical world, in our abstract experience of that world, and in the language we use to describe the physical and the abstract. Language is pervasive, and as a pervasive part of our ability to think and process the world, language lends itself as a ready tool for shaping the way people experience our designed worlds.

Schematic structures (Tool #1) provide a way to understand experience of concepts in the world. They help us form concepts to reason about the abstract world with concrete notions like direction, movement, containment, often represented with arrows, shapes, and lines. You can use schematic structures to plot basic narratives, flows of energy and sensory encounters.

Cognitive simulation (Tool #2) happens when some real world experience (e.g., motion) is simulated in the mind, providing a simulation framework that enables reasoning about the physical world or about abstract ideas. Cognitive simulation often involves schematic structures (e.g., directionality of motion) and involves attention patterns. Think of it as a moment of heightened awareness. Moments of cognitive simulation can serve to anchor an experience or to push the narrative of an experience forward. It is an emergent property of the organization of viewpoint.

Viewpoint (Tool #3) occurs with a major distinction between participant viewpoint and spectator viewpoint. Participant viewpoint is first-person and immersive, whereas spectator viewpoint is third-person and provides a vantage point for summary scanning. Viewpoint is different from perspective and serves as a framework for perspective. You can use viewpoint to shape a story, to create disorientation and orientation, or to build rhetorical effects into your design.

Embodiment (Tool #4) provides the basis for encountering and reasoning about the world. The body grounds our ability to reason about abstract ideas by relating them to our physical environment. It argues that we can think with our actions and that our bodies are part of our minds. If you can change something about the way someone experiences their body, you can change the way they think, because the body is a mode of thinking.

Motion (Tool #5) is a mode of being in the world. It is a skill that we use like an interface with the world around us. Motion gives structure to some of the ways that we reason about the world, and by intercepting those structures we can modulate experience along various dimensions, including orientation and disorientation, as well as viewpoint and attention. Motion also acts as a means of expression whether through meaning systems of gesture, body language, and dance, and is also, obviously, the way that people move through your designed environment.

The Senses (Tool #6) are the means we have for enacting **Perception**, and include visual (sight), auditory (sound), haptic (touch), olfactory (smell), and gustatory (taste) channels of sensory attention. Perception is the recognition of difference and variation in sensory data and we use it to identify meaning about the world and to understand symbolic content in the world as a relation of signal to noise. The senses oscillate data in figure-ground relationships as they are perceived. This helps us make sense out of space and time, influences memory, and contributes to emotional states and moods.

Emotions (Tool #7) are temporary experiences that we use to make meaning out of how we feel. They exist as salient elements of experience set against a pervasive background mood.

Some of these concepts might be new to you and you may be familiar with others. Take your time to let them sink in and refer back to these brief descriptions to clarify and remind you if you get confused. Ultimately, these concepts provide a set of tools that will help to structure the approaches you take during the design process of your experience. After working through these concepts you will be able to decide which ones are right for your project and which ones to save for later. These concepts will help you build your experience by enabling you to:

- choose a basic story and have a goal for your experience;
- provide your audience with a framework for inquiry into the sensory world;
- pair two or more sensory systems, viewpoints, conceptual models, physical systems, or elements of embodiment;
- couple that experiential pair with mental, emotional, or physical information;
- introduce a pattern of when and how specific sensory systems are active or activated;
- let your audience determine some element of their experience;
- create engagement points for triggers, feedback, openness, and bodily responses;
- use these engagement points to build your story or reach your goal;
- shape experience in the simplest ways for maximum effect;
- capture, focus, and direct attention through space, time, and information;
- use distraction appropriately by eliminating or creating it as it fits your story/goal;
- document your work as thoroughly as possible/necessary; and

— offer a memorable or insignificant experience (as determined by your desired goal).

Let's now look at these concepts in depth.

TOOL #1
Spatial Arrangement & Schematic Structures

Space is structured by the presence or absence of things that occupy space, define space, and provide avenues through space. These presences and absences do not just exist in the physical world, in fact they often become concrete building blocks for how we think about the abstract world (Lakoff and Johnson 1999; Tversky 2011). Before discussing how space is used in abstract reasoning, exploring the way that space is actively constructed helps identify some **basic spatial primitives**.

Lynch (1960) proposed a set of spatial primitives used in urban planning that also inform the spatial categorization of events, particularly events that feature path based motion and non-path based motion. His primitives are *node, path, edge, area,* and *landmark*. Nodes are points along a path, a path is the connection between nodes, edges are boundaries between areas, areas are the spaces contained within edges, and landmarks are a type of node that anchors an area and structures paths that cross edges to reach the landmark. In Chapter 9, this system of spatial organization will be discussed as part of a model for mapping an event space and categorizing areas in an event space. This will help to begin connecting spatial elements to activity/process elements when looking at the overall flow of the event experience.

These spatial arrangement tools may be familiar to you, but what might not be familiar is how they tie in with cognition. These primitives, along with the schemas below, show up in the way that people talk about space and reason about space. Consider the last time you gave someone directions. Likely you used a series of nodes, paths, landmarks, and other primitives to help scaffold the description in a memorable way.

Beyond giving people directions, schematic features show up in the metaphors people use to reason about the abstract world. Let's look at these schemas before offering an explanation.

Similar to the schematic nature of Lynch's spatial primitives, but arguably older and more pervasive in human cognition, image schemas provide a world-forming scaffold to artists and experience creators. Image schemas are pre-conceptual which means that they are the structure of concepts and show up in all types of objects, ideas, activities, beliefs, and other products of cognition (Figure 8).

We begin to experience these image schemas from birth (and possibly earlier in some cases) and they contribute to our ability to learn and adapt as we grow older. Think of image schemas as the skeletons of concepts. There are many schemas available to humans, but a basic list that suites the purposes of this book includes: **containment** (*container, in-out, full-empty, contained/contents*), **space** (*center-periphery, up-down, front-back, contact*), **plexity** (*unity-multiplicity, part-whole, link, collection, merging,* etc.), **motion** (*path, source-path-goal*), and **force** (*blockage, removal of restraint, compulsion,* etc.).

Looking at this list it is also obvious that these are not merely tools for organization in cognition, but that they also describe motion, composition, and energy in physical reality—capable of describing an earthquake's violent behavior (force, up-down, center-periphery), the way trees join together to make a forest (multiplex-uniplex, part-whole), the movement of a hunting cat (source-path-goal), or the way that syrup spreads out over a stack of pancakes (container, center-periphery). In fact, when these schemas are used in compositional strategies for 2D work they give a strong sense of dynamism in the imagery.

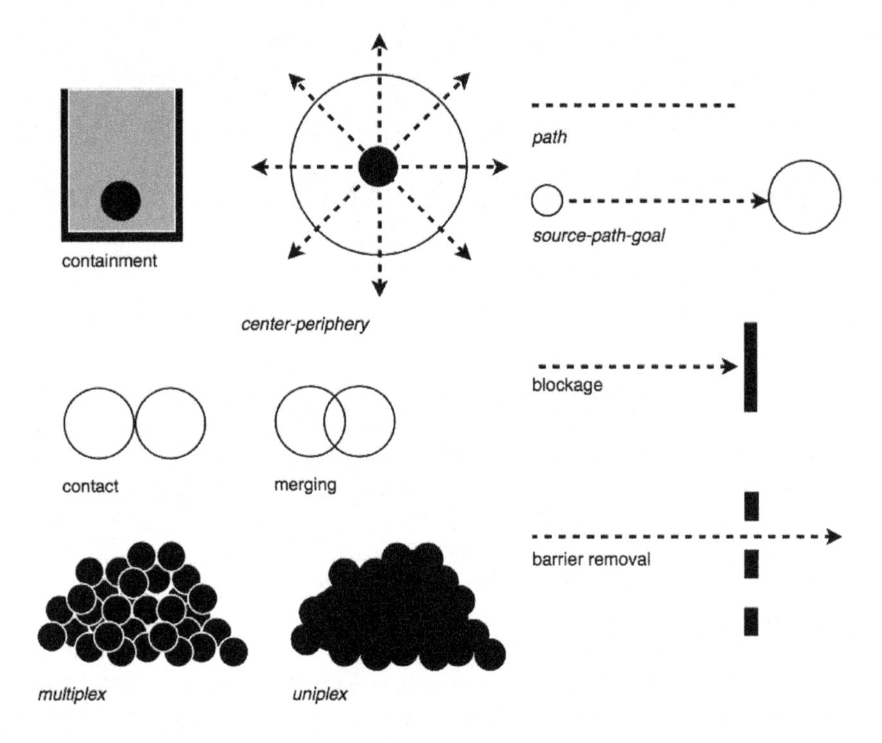

Figure 8. Basic Image Schemas.

Since concepts and physical conditions share image schematic properties there is a connection that forms between concrete and abstract reality—conceptual metaphors emerge to structure abstract concepts based on how image schemas structure our embodied experience of the world. Adopting conceptual metaphors in the design process of experiences provides a direct link between the mind and the world and provides source materials for novel conceptual scaffolding for people to make sense of the event experience.

As you will see throughout the remainder of this book, you can link abstract notions of things like *time, love,* or uncertainty to direct physical experience of things like *music, touch,* and spatial arrangement of *objects*.

Apply this now:

— Start looking at daily life in terms of these schema to get a sense for how they exist in the wild. Think about ways to borrow the natural occurrence of these schema in your own work.
— Start looking at your art in terms of what schemas you already use. If you paint, you might notice them in the directionality of the brush strokes. But if you are strictly an installation artist, find the flows of energy and containment and clustering in your installations. Do those flows follow a schema? If not, could they?
— Start looking at other people's art in terms of these schemas. Which schemas do the more cohesive works of art use most often? Does the artist routinely work with the same schema? How does the schema structure the content and subject of the work?

The Sculptural Qualities of Spatial Events

Think about installations like a sculpture that is turned into a spatial event. The sculpture has an armature on which material is formed, and that material communicates the content of the sculpture. The visual qualities of the content tell something about the sculpture, but you can also tell something about the sculpture with the materials and techniques you choose to use. It's the same way with installations, and armatures in this case are the schematic structures of the installation, the skeleton of the experience.

Determine what skeletal structure your experience will have and then fill it in with the form and content. Sometimes your armature will be the pathway segments and you will fill in the form and content along the path segments. Sometimes your armature will be the nodes and you will fill in the form and content at the nodes, leaving the path between the nodes unarticulated, formless and lacking in content. Sometimes your armature will be the pathway *and* the nodes along the path, in which case you fill in content for everything. Sometimes with the most complex installations you will have the armature be an abstract experience that finds its form and content in the expression of the path. In this case, the path acts as the physical anchor for an abstract concept, for instance, the idea of *Life is a Journey* is a metaphor system in which the content (in this case, the departure-to-destination experience of life) is reified in the concrete terms of a journey (like paths and roads).

How to Use Schemas to Structure an Experience: Examples of Source-Path-Goal, Paths, Containers, and Narratives

Schemas such as the Source-Path-Goal schema fit nicely with narrative structure. *Life Is a Journey* is a particular conceptual metaphor (a metaphor that uses our physical experience to structure our abstract experience) that uses the Source-Path-Goal schema. More on conceptual metaphors in the section on embodiment (Tool #4) and the chapter on narrative (Chapter 5), for now let's look briefly at how this schema helps tell a story.

Path Schemas and Narratives as Design Primitives for Experience Design

When you design a story, design it as a path. Narrative arcs can be plotted with the rule of thirds grid to translate a story into a physical experience. This makes the narrative into a kind of physical path. Instead of following a story line by reading it, visitors follow a story line by walking along the story line as a path. This makes use of the physical metaphor: *Story/Experience is a Path* (a variant of *Life is a Journey*), and it makes use of the schema: *Paths connect Places* (a variant of the *Source-Path-Goal* schema).

This schema *Paths connect Places* is easy to grasp because it is such a common experience in our everyday lives. If you want to move from Point A to Point B, you take a path between them. A and B are connected by a path. All locations, nodes, landmarks and spots connect through some path. The path can be *ad hoc,* where someone chooses a path that is not predetermined, or the path can be determined and people have to take the path that is already decided for them. *Ad hoc* paths are paths that users choose, while *determined* paths are paths that the designer chooses. The story with each type is clearly different, the user drives the story in an *ad hoc* path (while working with the limitations in the universe of content that the designer created), while the designer drives the story in the determined path. Both can be useful and interesting.

A path defines the experience that a user has by providing different information to the user. Some paths provide viewpoint information, mixing up immersive participant views with the vistas and views of a spectator (more about this in Tool #3). Other paths focus on using participant and spectator perspectives with the non-visual senses. Still other paths can provide information that relates to time and temporal experience, and others provide direct information through text and language (known in the literary world as a sequence of "information dumps"). A path *reveals* information as the visitor walks along the path and comes across the different intervention nodes in the space. The segments of the path that connect the nodes can be any length, and the lengths of path combined with the frequency and spacing of the nodes creates a kind of rhythm to the experience of walking along the path. For example, if you want to borrow the rule of thirds into your path design, and you want to coordinate that with the climax of your narrative, you would design a path that has the climax node in the third segment of a four segment path (or whatever multiple of 4 that structures your path).

Paths that focus on visual experience (such as an outdoor path focused on alternating views of a landscape) define the experience of a participant by providing different types of views in some sequence. The path will oscillate between immersive participant views and removed spectator views and vistas at whatever rate and frequency that you as the path designer/story teller decide.

Scale is another dimension of paths. Paths can be designed on a small scale (such as the path you want someone's eye to follow in an image) or they can be designed on larger scales, such as paths through a building, through an environment, or through a forest. Paths take different lengths of time to follow. Some paths are intended to be completed in a single short session while others can require repeat visits to experience all of the details (think about how some museum exhibits always seem to have something you didn't see before). There is no reason that you can't design paths that take years or even lifetimes to unfold completely. While conventional gallery experiences may only last an afternoon, nothing prevents artists from designing path experiences that are much longer.

Path linearity can be coordinated with the linearity of your narrative. A linear narrative will have a rather straightforward linear path, whereas a non-linear narrative might not have a determined path or might provide an unstructured path or general ambient environment. A narrative that is multi-linear will have different narrative paths that, depending on which path you take, your experience follows one narrative out of many alternative options (the *Choose Your Own Adventure* series of books is a great example of this). Linearity in path shape is the ordered sequence of the path linked with some narrative. If the experience is non-narrative you could still use a linear path without using a narrative to tie it all together. This type of experience would have the feel of assemblage, *bricolage,* and even randomness, perhaps evoking a surrealist dream-like experience. This is not to say that primal and visceral experience can only be built into a non-narrative experience, instead, these emotive and visceral sensations can be applied to any of the path shapes.

How do you give people the experience of ownership in an experience?

Path entry points provide access into the experience, and the entry point determines the part of the story a person is walking in on. An entry point can be a physical entry point like a door, or it can be a sensory entry point like a sound that the visitor hears in the background that draws them in toward your designed experience. "Entry point" is more of a concept and how it manifests in your design is up to you.

Keep in mind that multiple entry points means that people will not all enter the experience with the same background information. People entering an experience at points later than others will not have learned the story world in the same way. You may have to catch new people up to speed with the story, or perhaps you want to keep new people in the dark. Whether or not you catch people up to speed with the earlier parts of the narrative that they have missed, you should help people acquire a sense of ownership or belonging in the experience to make your visitor feel comfortable enough to stay in the scene without simply exiting the experience.

In this kind of experience where people enter from multiple entry points, you could say it has a metaphor schema of something like: *Multiple Paths lead to a Unified Story,* combined with the spatial metaphor: *Experiences are Containers.*

If you look at a map of a river system, the smaller streams and brooks and creeks that feed it resemble the system of arteries and veins in a body. Borrowing from this dendritic pattern, the narrative path schema of your designed experience is like the main body of the river. The smaller flows are the series of paths that lead into the main thrusting narrative path. Your goal is to funnel all of the smaller flows into the main experience of your work. Funnel all attention and use the smaller flowing streams to catch people by surprise and lead them into the main experience.

If you hike through the mountains and come across a small stream and you follow it in the direction it flows, you will eventually come to a large river which if followed long enough takes you to the ocean. The experience as a participant is linear, a path schema, and you only see what lays in front of you. The other tributaries that feed the river are invisible to you, blocked by the contours of the terrain jumping up around you. But that doesn't change your experience of the main river. You experience your path to the main experience and you enter the main experience with your path-based conditioning.

One way to open people to the experience is to give them something that is familiar that they can identify and use to make sense of the experience as they enter into the experience. One way to do this might be to use the language of doors and windows to help people think about the experience as a building or a room. Doors, doorways, thresholds, stairs, windows, and even screens give a sense of in and out. You often walk *into* a building and look *out* the window. This means that you can think of a room or a building as a type of container that has an inside and an outside. You can extend this idea to other environments like cities, neighborhoods, galleries, forests, and so on. All of these forms have *insides* and *outsides,* and borrowing doors (by physically placing doorways along a path, or by using language that suggests a doorway) can help you provide an entry point that is conceptually recognizable as an entrance into an experience.

We think about states, locations, events, and emotions as types of containers that we move between as we go about our daily lives. "I'm **in** a bad mood, and I have to go **into** a meeting but I'd rather get **out** of here," or "She's **in** a better place now that she left **from** that relationship." These aren't just figures of speech, they are ways that people think about abstract mental states by using a concrete notion of containment, and English happens to point this out by the way that we use prepositions. Since experience in general is a container, where you place the "openings" to your controlled experience depends on how you want people to engage. Experiences often have literal openings because the experience takes place in a discrete location (e.g., *in a room* or *in the woods*) or in some other spatial container, but they also have less tangible openings because experiences take place in a time frame, in a context, in a particular order, in no particular order.

Learn to think about experiences as containers for controlled events. These experience containers need an opening for people to enter the experience.

People go *into* the experience container and come *out* with memories and lingering sensations, new knowledge states, and hopefully new emotional states. The experience container acts very much like a reaction chamber where energy is fed into the chamber in one state and converts to a new state before exiting. A change takes place in these container-reactors.

Learn to think about experiences as container for events and memories and sensations. Experiences are like containers that people enter into and so you need to provide a door into the experience. This doorway is important because it also acts as an exit out of a person's everyday experience as they enter into your engineered experience. Entry points are transitions and they are the first transformational experience as people put aside their everyday and enter into the new experience you are providing. Perhaps your entry point is a physical location like a tunnel into a space or an actual door, but equally it can be a sensory entry point like a sudden waft of a haunting and familiar smell. You might also use a training area as the entry point where people are presented with signs and symbols and other stimuli that prepare and prime them in some way for the experience in which they are about to partake. Maybe a little classroom where patrons sit for a lecture or briefing acts as the door. Whatever your door is, it needs to help people exit their distractions and enter a world of focused attention on whatever it is you want them to focus.

These notions of path schema and narrative are covered more extensively throughout this book, as are the notions of sequencing, viewpoints, and conceptual metaphors.

TOOL #2

Cognitive Simulation

Cognitive simulation is a tool that you can use to build your experience by the combination of physical forms, sensory layers, or visual fields. It evokes a sort of dynamic sensation to the experience by activating the experience with mental imagery. You probably already know how to produce a type of stimuli that sometimes activates a form of cognitive simulation called fictive motion if you know how to use perspective in drawing (Image 1).

Think about perspective drawing and leading lines that pull you into the depth of a two-dimensional image, your attention scans along the trajectory of the perspective lines as you move into the image even though it is a flat surface. If you ask someone to describe an image that uses perspective drawing they will likely use motion words to describe the line of sight: "the road runs into the distance" or "that fence follows the tree line." Images that use perspective effectively have more dynamism than images that fail to use perspective correctly. We call those failures flat images and describe them as static and unidimensional because the composition does not encourage the eye to shift in space.

Certain object shapes provide better paths for your eye to scan along than others. For instance, if you want a drawing to evoke a sense of distance, you would include long linear path shapes that span between two points. The long lines will pull the eye, and pull the focal point of attention along that line in an act of mental simulation in which your attention moves along the trajectory path provided by the long line. We will expand on this a little later on in the text.

These two dimensional techniques work because they mimic the way that dimensionality and perspective work in real life. We translate our physical concrete experience of the world into the techniques we use for the two dimensional abstraction of the world as we produce images. With this in mind, we can readily apply the compositional techniques from two dimensional art to three dimensional space. Now let's bring in the idea of cognitive simulation.

Instead of thinking about the eye scanning along a line in an image or environment, reframe it as attention scanning along the line shape (Matlock 2004a; Matlock 2004b; Oakley 2009; Talmy 2001). This is what happens when you hear a sentence like "this stone wall runs right through the forest and ends at the river." You simulate motion along a linear trajectory as your attention scans along the mental image of the stone wall (or possibly just a schematic line) extending between the beginning of the wall and the end point of the wall. In fact, this attentional scanning defines the type of cognitive simulation that scientists call fictive motion. Fictive motion acts in the mind like perspective does in a drawing: it animates the static world and creates a dynamic scene.

Learning how to arrange the physical world in such a way that elicits fictive motion will help you in designing experiences that evoke responses because it provides a direct doorway into attention through mental simulation. It might be possible to achieve this effect by providing people with both participant and spectator perspective as they move through the environment. Let's briefly look at how two types of viewpoint, participant perspective and spectator perspective, give us different information about a visual scene and see how that might influence cognitive simulation in an engineered environment (for a more detailed description of viewpoint, see Tool #3).

Images that have dynamic sensations have clear linear shapes for the eye to follow. Think of this linearity as path shape. When

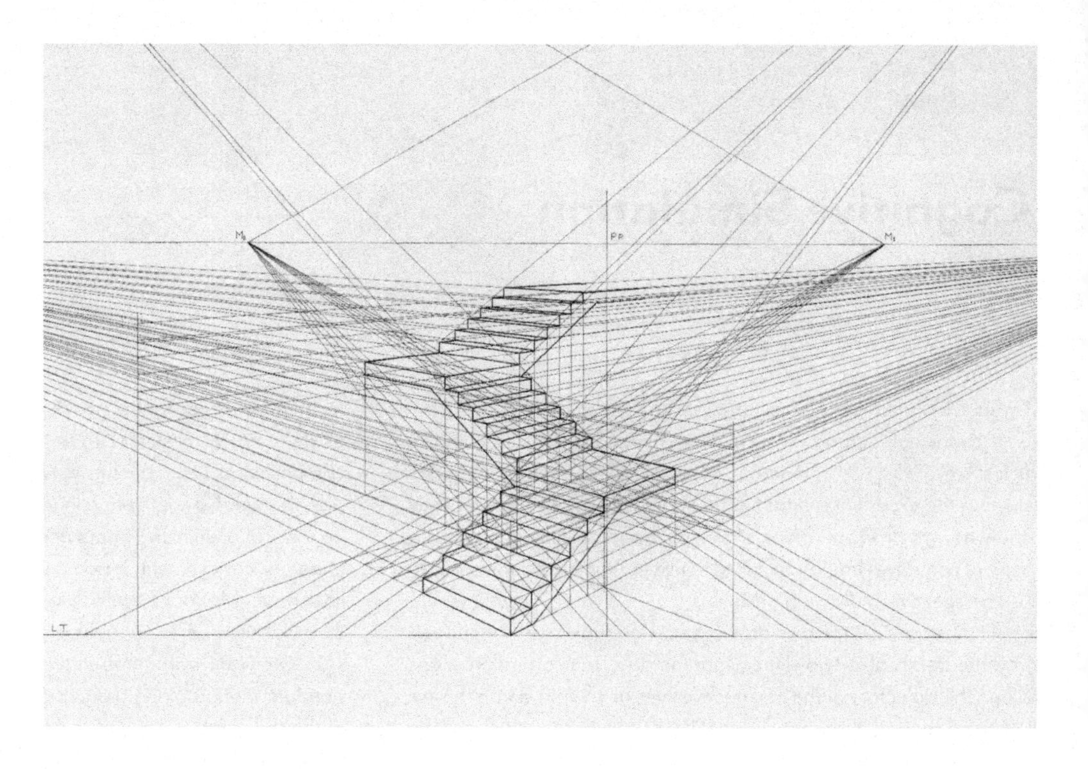

the eye can see the path shape it can follow the path shape. Think of the ability to follow path shape as a form of spectatorship, or third person perspective, or distance. You gain this spectator viewpoint from aerial views, from maps, from balconies and windows, from views at a distance. Spectatorship gives you vertical, high-level summary information about a visual scene.

When the eye cannot see the path shape it cannot follow the path shape. Think of this inability to visually follow path shape as resulting from not taking the spectator viewpoint but instead engaging in participation, first person perspective, or immersion. You take this participant viewpoint when your environment is consuming, swallowing, and immersive such as when walking at street level or following a path in the forest. Participation gives you horizontal, low-level detailed information about a visual scene.

Viewpoint may influence the way that people use cognitive simulation. In a study on movement along forest paths (Dewey 2012), viewpoint seemed to influence the way that people used fictive motion in their descriptions of the path. People who walked along the forest path without a map of the path could only see the path ahead of them. People who walked along the forest path with a map of the path (Image 2) could see both the path ahead of them and the overall path shape as a line on their map. People who had access to the overall path shape described the path by using fictive motion more than the people without maps who could only see the path ahead. This seems to suggest that having spectator viewpoint (i.e., the ability to see path shape) enables people to use attentional scanning to make sense out of their participant viewpoint experience of the physical world. Cognitive simulation is at play in the ways that people use viewpoint to understand their surroundings.

It is difficult to discern the overall path shape of a trail that zigzags through hilly forest. The constant change of direction along the vertical axis when moving up and down hillsides, combined with the constant change of direction on the horizontal axis at the trail switchbacks, make it difficult to perceive the path shape while walk-

Image 1. By Luciano Testoni (1995). Wikimedia Commons.

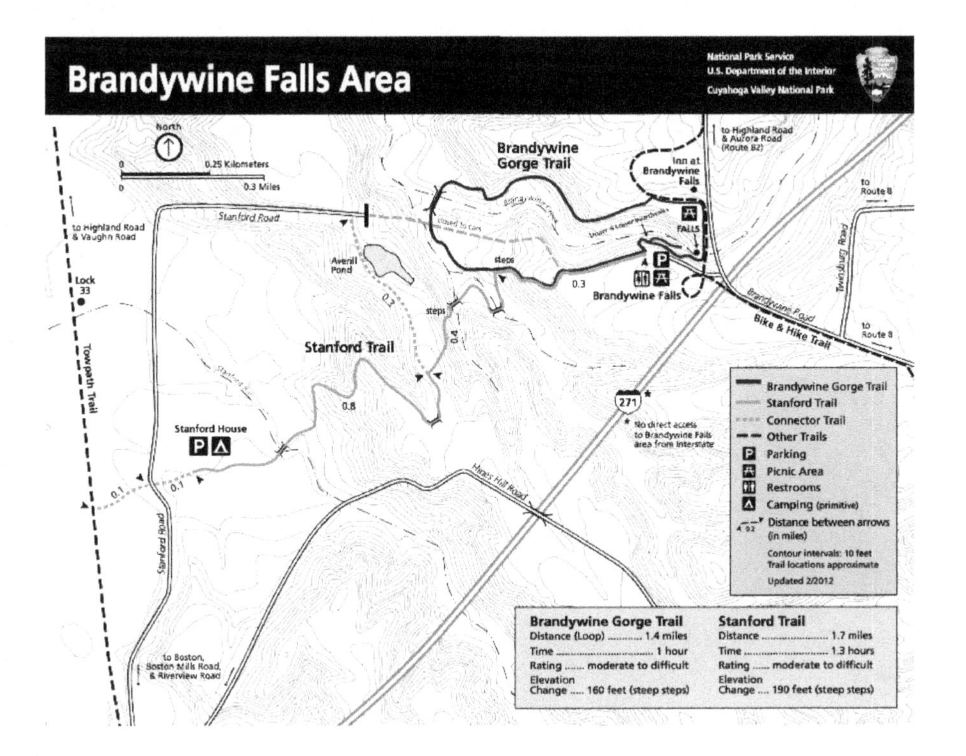

ing on the trail. But if you look at the plan view (a.k.a. the bird's eye view, or the map view), it is easy to perceive the path shape as a zigzagging shape.

If viewpoint facilitates the kind of cognitive simulation found in fictive motion, then carefully combining spectator and participant viewpoints can create environments that evoke active and dynamic attentional scanning, making the environment seem to come alive. Next we look in depth at viewpoint and the way that it can be used to structure experience.

Image 2. Brandywine Falls area map, featuring the Stanford Trail, Cuyahoga Valley National Park. National Parks Service, U.S. Department of the Interior.

TOOL #3
Viewpoint

The way that we see the world either gives us information or takes information away from us. If you are a spectator you see more and get extra information that is unavailable to participants who can't see the big picture, but it also means that as a spectator you don't have all of the details that the participant has, you only have what you can observe from afar. Participants do not have the privileged perspective of the spectator, but they do see how things are on the ground. This participant-spectator separation of viewpoint is a foundational concept in experience and can be manipulated in ways that can enhance an experience or augment existing awareness. It can be used to disorient or to bring clarity, both of which can shape experiences in positive and negative ways. It is also possible to have both viewpoints, a blend of participant viewpoint and spectator viewpoint, increasing people's capacity to make informed decisions. Consider the role of satellite navigation; a driver experiences the road from a participant viewpoint, but also benefits from the spectator viewpoint of a dynamic map of the road. As a way of breaking up experience of the world, viewpoint becomes a basic tool for cognitive engineering. It is also linked to increased use of cognitive simulation (Dewey 2012) and can be used to orient or disorient viewers through oscillation and sequencing (Dewey 2014).

In this section we continue the discussion of viewpoint introduced in the preceding section and explore how environment can influence spectator viewpoint and participant viewpoint, and how both forms of viewpoint can be combined to create rhetorical effects in the engineered experience. To review the definitions provided earlier, spectator viewpoint takes third person perspective, it is removed and distant, giving a near vertical view that provides a summary view of a scene that the spectator can scan from afar. Participant viewpoint is first person perspective, it is immersed and horizontal because the participant is in the details of the visual scene.

Control Viewpoint to Build Perspective-Based Experiences of Meaning

Participant and spectator viewpoint show up at all levels of the experience which means that playing with viewpoint makes it possible to change an experience. When you successfully modulate viewpoint it gives people new perspectives on their experience.

When we experience the world as participants we engage in a horizontal world of experience—participant views are immersive because they cast the surrounding environment in a kind of metaphorical container for experience and the participant is *inside* the experience. Spectator views are not immersive, instead they are observing states during which the *outside* spectator looks *into* the container to see what is happening to the participant. Sometimes both participant and spectator roles can be blended together (the above mentioned satellite navigation), this is a special both/and view of your location and movement in space—you see the horizontal view of the city and the schematic vertical view of the city on the map. This blended role increases your ability to orient yourself. In other cases, oscillating between the two viewpoints can give or take away knowledge about the subject: bouncing between participant and spectator can help you track what is happening in an experience, but it can also be used to confuse and disorient you. These types of viewpoint control are crucial tools for hacking experience.

Bringing viewpoint into your experience design is kind of like telling a story. Novels told in an omniscient third person voice often use both viewpoints (both inner subjective thoughts of characters and the external perspective of the narrator). You can take a similar approach in your installation by playing with how much and what type of viewpoint information they have access to at any particular point in the experience. You could achieve this by writing a script for the experience that you want to design. It might follow a pattern like this:

A person enters the designed experience without knowing what lies ahead. As they move along you have selected to control the viewpoints that are available to them. Control the experience so that sometimes the person can only experience a participant-immersive viewpoint, and at other times they can only experience a spectator viewpoint. Switch back and forth between the two types of viewpoints to give the person a fragmented perspective of what they are experiencing, like different angles of a fragmented story. Then slowly begin to combine viewpoints so that the person gets a blended participant and spectator viewpoint as details begin to fall into place and make sense with some central, driving thrust. At this point when the viewpoints blend, the visitor may experience a moment rich with cognitive simulation, a point of high alertness to the fabric of the experience, or even a recognition of the experience as a story. This moment of alertness is where you communicate to the viewer any content messages you want them to take away from the experience because this moment is full of rhetorical effect. You could think about this moment like the climax of a narrative arc, the place where all of the context and rising action culminate and tie the experience together. Slowly disintegrate and untangle the participant and spectator viewpoints and fade out with a modulation of the participant and spectator viewpoints until there is no perceivable viewpoint but only an impressionistic memory of the overall experience.

If you think about designing experiences that are like stories in physical space, using viewpoint to tell those stories doesn't just have to be visual. Viewpoint is a part of all of the senses, and just like blended moments of spectator and participant viewpoint, blends of viewpoint that involve multiple senses create dynamic moments.

Multi-Sensory Approaches to Viewpoint

A participant viewpoint and a spectator viewpoint can belong to different sensory systems. Maybe in the beginning of the experience you want the person to experience visual information as a participant and you want them to experience auditory information as an eavesdropping spectator. The auditory cues or stimuli belong to a different scene than the visual cues. As you begin to blend the auditory spectator viewpoint with the visual participant viewpoint you begin to bring both sound and image into the same scene. Maybe as the person moves in the space the visual scene becomes the scene where the sound is being generated from the spectator viewpoint. Or maybe as the person moves in the space the spectator sound fades away and other sounds begin to match the visual participant channel. This happens in real life all the time, we hear something that we don't see because we are looking at something else, or you are eating food that has a distinct smell (you experience the food smell as a participant) and someone walks in with perfume that overpowers your sense of smell (you experience the perfume smell as a spectator).

Conflicting Sensory Viewpoints and Increased Mental Processing

Because these multi-sensory experiences of viewpoint are so banal they make good places to engage in hacking, especially if spectator and participant both start off coordinated with each other and then begin to diverge in an increasingly dissonant manner. Processing sensory information that is in harmony is easier than processing information that is contradictory. For instance, with language processing, the Stroop test is a reaction time test that shows that it is harder to read the name of a color out loud when the text is printed in a color that is different from the name of the color (so, for example, the color name *black* printed in *blue*). This applies to more than colors, and reaction time effects like the Action-Sentence Compatibility Effect (Glenberg and Kaschak 2002; Bergen and Wheeler 2010; Sato et al. 2013) show that when a subject is performing an action that is compatible with a sentence they are reading that their completion of the action is faster, but slower when the action and the sentence are not compatible, representing an increase in the difficulty of processing the stimuli. By bringing together sensory stimuli that have conflicting viewpoints, similar compatibility effects can be achieved, giving conflicting texture to the experience.

You might have a participant viewpoint that is strictly auditory and a spectator viewpoint that is strictly visual. Or perhaps your participant viewpoint starts out in the auditory channel but becomes visual or olfactory. Maybe your participant viewpoint and your spectator viewpoint work with the same sensory channel, both of them being auditory channels, but with different levels of scale or granularity or distance. These sensory qualities are discussed more in the text and tables in Tool #6. But for now, it is important to understand that you can control viewpoint so that it makes use of particular senses and that the flow of viewpoint throughout an experience is what causes the sensory data to evoke cognitive simulation. In order to create an engaging experience, viewpoint can't be a static element.

Viewpoint is dynamic and oscillates and changes over time as the environment changes and our attention shifts. Viewpoint is an extension of attention, in the same way that attention drives the shifting figure in the perception of figure-ground organization, attention also drives the oscillation of viewpoint.

Think about the act of climbing a set of stairs, along the climb you have the participant viewpoint of the path up the stairwell. Once you reach the top you get a spectator viewpoint either out of a window, or you get the spectator viewpoint of the stairs below.

Or, maybe you are walking down a hall in a museum and you see the gallery up ahead and you can see art on the walls and patrons and you have a spectator view of what is going on inside the gallery at the same time that you have participant view of the hall. Once you cross the threshold of the gallery doorway you take a participant view in that space. If you turned around and looked down the hall you would now have the spectator view of the hall.

Or, you are sitting in a room immersed in participant view and you turn to look out the window, becoming a spectator. Similarly, when you sit in a car and have awareness of the interior of the car, you are immersed in the vehicle with participant view. Looking out the window as you move through the landscape gives you spectator viewpoint. Or at a meal, you are participant immersed in table conversation in the participant context of a dining room, and then you look down to decide what your next bite will be and you take the spectator view of the food on your plate, macro-participant to micro-spectator.

These banal examples demonstrate how pervasive the shift between viewpoints is in our daily lives, and motion and scale and attention all facilitate this shift. Because this is such a pervasive element of life, it provides a subtle portal for alerting experience.

Some environments (e.g., a room with windows) make it easier to take both viewpoints while others restrict viewpoints (e.g., a tunnel). These affordances and restrictions can be more than architectural. Because the participant view is immersive, simply overwhelming the senses with intense, layered, or focused sensory stimuli might prompt a feeling of immersion. When designing an experience, decide which elements of the experience will shift viewpoint for participants and which elements will restrict and reinforce viewpoint. Test them out with a group of friends, or simply find a scenario in everyday life (such as walking up the stairs) and translate the viewpoint shifts into your designed interventions in the gallery.

By placing the shift of viewpoint along a path experience, you can begin to harness the oscillation of viewpoint to create rhetorical effects along the path. This link of perceptual content to spatial structure can also map to narrative content (see Chapters 5 and 6).

Spectators *observe* a scene while participants *engage* in a scene. This seems simple enough. But there are distinct advantages to having one view or the other. Spectators can see what is going on at a higher level while participants experience lower-level decision making. Spectator views often occur on vertical axes while participant views often occur on horizontal axes. This is most clear when you think about maps and travel through the countryside. You have a vertical view of the layout of the land with the map, and you have a horizontal view of the layout of the land as you drive through it. By keeping track of where you are on the map you gain a spectator viewpoint of your location while also keeping your participant viewpoint as a traveler. Your satellite navigation automates that for you, but the principle is the same.

This distinction between spectator and participant becomes useful as a rhetorical tool. It can be used to give privileged views to bring clarity and orientation to a scene or it can be used to take away information bringing disorientation. As a designer, using both viewpoints in a sequence can be useful for achieving atmospheric moods and environmental qualities like confusion, remoteness, refuge, isolation. Viewpoints can be used to create momentary confusion or sustained confusion.

This distinction between spectator and participant also becomes useful as a narrative device. If spectator viewpoint is top-down and participant viewpoint is bottom up, then in a narrative intervention which features constant oscillations between spectator and participant viewpoint in an ad hoc manner, the narrative effect can be thought of as an object-oriented narrative and the story that the viewer experiences emerges as a particular selection of narrative threads based on what the viewer wants to see and how the viewer actually moves through the intervention space. That means that for each viewer, the record of their movement through the space can be read as a kind of index to what they find to be of personal visual interest. It also means that the designer of the experience can tell a unilinear story, a multi-linear story, or a non-linear story.

Oscillation of Viewpoint as a Rhetorical Tool

Viewpoint oscillation also works well in curatorial practice to give a narrative to an exhibit. Chapter 5 presents a case study of a photography exhibit in an art museum that explored the role of spectator and participant viewpoint in creating a textured story about the sustained devastation in the decade following the 1980 eruption of Mount St. Helens. Basic oscillations between photographs that featured spectator views of the volcanic landscape and photographs that featured participant views on the ground in various locations created a sense of motion through the exhibit and created a feeling of disorientation which was one of the rhetorical goals of the exhibit. It also felt cinematic as if the switch from participant viewpoint images to spectator images were camera cuts giving different angles on the scene. Sometimes a group of participant images would have no spectator images to give context to that participant view. In those cases, this increased a feeling of isolation by removing any chance of establishing spatial coordinates in the story world of the photographs. Oscillation between viewpoints might seem like a simple device, but it gives enormous variation in the fabric of a story. Each oscillation contributes a new perspective on information and this accumulates as the viewpoint continues to oscillate until the collection of viewpoints on a scene tell their own story. This accretion of different viewpoints structured the story of the photography exhibit, but it happens in any experience where viewpoint oscillates. You gradually gain more insight into a problem, or a setting, or a person's character. Every switch in viewpoint provides another facet of the story you are experiencing.

Consider the way that a film is a collection of still images set in motion by the sequential viewing of each frame. As mentioned before (in Chapters 2 and 3), this is also a way to break down motion in daily life as well as motion in the designed world of your installation. Just like the Rule of Thirds applies to each still in a film and each snapshot moment in daily life, and each snapshot moment in your installations, so to this succession and oscillation of viewpoint make up the cinematic and oceanic feel of both the fabric of life and the feel of the visitor movement through a designed intervention in the experience space.

Blended Viewpoints and Path Schema

In all situations you will at least have one perspective or the other, when you have both perspectives it is a privileged position, and your brain recognizes this privilege too. In fact, having both perspectives seems to facilitate cognitive simulation which helps you make sense out of the world around you. In the experimental study of people walking in the woods (discussed in Tool #2), being in a spectator viewpoint situation but having extra access to the spectator viewpoint (with a map) gave people a richer knowledge of their surroundings and the blended viewpoints helped them reason about their immediate environment so that they could make decisions about what actions to take along their journey. The people who only had a participant perspective described visual environments as static, but people with both perspectives described the same visual scenes as dynamic environments and were able to use clues in the environment to make sense out of their location in that environment. This is interesting because it shows that providing a path without a map can be immersive without necessarily being viewed as dynamic (from the perspective of cognitive simulation).

People who only had a participant immersive viewpoint describe their own motion through the environment rather also using descriptions which construed the environment itself in active terms. They operated with a limited ***path** image schema*. In that experiment, people without maps described their own motion along the trail because that is all they could see, but people who had a map had a bird's eye view and they described the trail as if it were in motion (e.g., *the trail winds around, the trail goes this way, the trail curves ahead*).

People who had access to spectator viewpoint (through their map) used rich descriptions of path shape in their motion language about the trail. This indicates that their conceptualization of space included large scale motion. Having the spectator view enabled them to scan along the entire length of the trail on the map and to see all of the directions and shapes formed by the trail. They saw the beginning of the trail, the end of the trail, and everything in between and were able to accurately describe the path that the trail took through the woods. They made use of the ***source-path-goal** image schema* (as opposed to the simple path schema of participant viewpoint). The source-path-goal (s-p-g) schema (illustrated in Tool #1 and in Figure 8) is a schema that is typically available in blended viewpoints, and people who used the s-p-g schema in this study accurately described their experience with the confidence of knowing what was happening around them. The people who only had participant viewpoint, on the other hand, gave poor descriptions of path shape when they talked about the trail. They could not scan the length of the trail, and so they had no idea where the trail was leading or what the overall trail looked like. They had limited visibility and only saw what was in front of them. They only had the *path* schema. The point here is that only giving your audience an immersive participant viewpoint will prevent them from being able to anticipate, remember, and describe the shape of the path. In other words, it doesn't let them track where they are on the path when they are experiencing certain elements of the path. This could be very useful if your design needs to confuse or disorient people in a simple way. Although giving your audience multiple viewpoints can also achieve this disorienting effect (in some cases more powerfully) if your audience has an immersive participant viewpoint and every once in a while you give them a misleading spectator viewpoint.

In a complex environment like a forest where there is limited visibility because of the trees and terrain, a line of sight *path* schema severely limits environmental awareness, it contributes to a feeling of isolation (and either the positive or negative emotional connotations of that feeling), it might lead to uncertainty and anticipation about what is around the next bend, it can lead to feelings of either security or vulnerability (both of which can be false impressions), and a path schema can result in feelings of both aimlessness or adventure because the destination/goal is unknown. You can achieve a *path* schema in your work by employing participant viewpoint as a means of disorienting people and keeping them from seeing the bigger picture.

Building a *source-path-goal* schema into your experience helps people comprehend the environment, task, activity, or process they are engaging in because it lets them see the beginning, end, and steps along the way. While a *path* schema limits awareness, a *source-path-goal* schema increases awareness of the overall trajectory, destination, and purpose of a path. In the example of the limited visibility of the forest, a source-path-goal schema, as provided by a map of a path, brings clarity and increases what a person can experience of the forest.

This discussion of schema and viewpoint is useful for both galleries and artists designing installations because it provides places to tell more of the story. It helps you decide where to place pieces (whether they are didactic pieces, sculptural pieces, or new elements of the installation) for stronger emotional cueing. It helps you to build a path for sensory inquiry that creates an experience of cumulative sensing by layering sensory viewpoints (like spectator sounds with participant images, etc.).

Viewpoint and Learning

Viewpoint gives different experiences to spectator and participant. Choreographer and dancer Yvonne Rainer (2006) discusses how she notices differences between dancers who have learned her choreography from watching filmed versions of her work compared to dancers who learned from a live choreographer. The differences show up in the performances. This difference comes from a difference in viewpoint. She mentions that the dimensionality of the dance marked out (or traced with the body) by an instructor in lines on the floor is not able to be attained by those viewing the film of the performance. This is a difference between participant and spectator, the spectator view only approximates her work, but the participant view enables acquisition and accurate dissemination of the choreography of the dance. This is similar to the English translation of the nuanced Chinese Confucian statement: "I hear and I forget. I see and I remember. I do and I understand," where *seeing* is spectatorship and *doing* is participation. Learning can't just be a process of observing, or spectatorship because it doesn't ask the body to prove that it comprehends what it has observed. When the body is performing tasks of manual literacy such as dance or other controlled practices, it is *enactive* and operating in participant viewpoint. Bring this concept into experience design by designing interventions and installations that require participation rather than mere spectatorship.

Reveal Through Movement

Design an experience that can't be observed unless the audience participates in making that experience happen; aim to have participation animate the experience and make the act of participation a reflective process to blend participant and spectator roles in a conceptual manner. Get people to think-by-acting by producing an experience that is *inquiry-based* in some sensory category where people have to actually do little tasks like smell certain combinations of scents, or find and follow a scent path. Make them reflect and become emotionally introspective by using sensory stimuli that are linked to other stimuli (the smell of the sea is linked to the saltiness of the air, the humidity of the shore, the sound of waves, the warmth of the sun, and the texture of the sand). Many people have positive memories of beaches, and stimuli that suggests a beach might trigger positive memories, while for other people the beach is an awful place and it might trigger negative or neutral memories. Either way, you are suggesting a place by replicating stimuli and letting people figure out what is going on by moving through the experience, accumulating clues, and thinking about the place in a way that will likely trigger emotional content.

Using the thinking-by-acting model, you could present a set of scents that fill two different zones in an environment and prompt people with introspective questions as they enter each new scent zone in that environment. Your goal is not so much to program an emotional response to a particular scent, but to present a sequence of scents and let them trigger different memories in the audience as the audience members move through the space. You provide two different prompts: one is the *control* (the scent) and the other is uniform (the introspective question) but *variable* (each person's answer to the question is different). You can even sequence the prompts

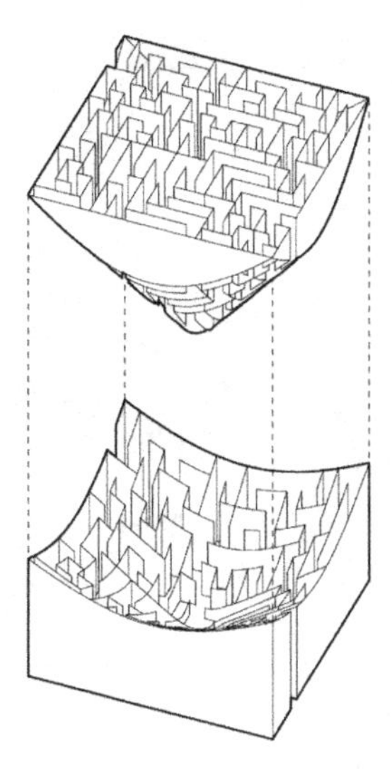

EXCAVATE CONE OF VISION
Removal of material within the visitor's cone of vision offers visibility.

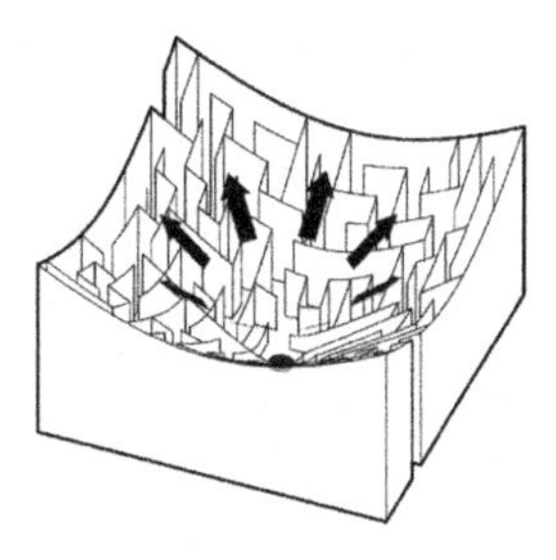

THE FINAL REVEAL
The Maze becomes a panopticon that displays the interior in its entirety!

Image 3. *Excavate Cone of Vision.* Courtesy of BIG – Bjarke Ingels Group.
Image 4. *The Final Reveal.* Courtesy of BIG – Bjarke Ingels Group.

and get a cumulative effect, and you can return to earlier prompts throughout the duration of the sequence to shift and reframe perception and memory in the experience. A scent-filled room and the simple prompt question "how does **this** make you feel?" directs attention to the scent (an attention element) and blends immersion (a physical element) with introspection (a memory element) to get at an emotional response.

Participant/Spectator Effects while Moving Through Spaces

Wayfinding and navigation are two different modes of moving through space. Navigation relies on absolute directions and maps of an area. Navigators plot movement through the area by looking for optimal routes and decide on turn-by-turn travel before they encounter the obstacles they are avoiding. Navigation is largely dependent upon spatial coordinates and is similar to spectator viewpoint. Wayfinding, on the other hand, is a process of encountering obstacles and finding your way around them on the spot, or encountering intersections and deciding which way to turn based on landmarks. Wayfinding is largely dependent upon landmarks and the visual field and is similar to participant viewpoint. Vision is a critical tool that people use in wayfinding in unfamiliar environments. It makes sense, then, that removing or manipulating visual information would disorient people in an unfamiliar environment. This comes in handy for designing experiences that quickly disrupt audience confidence, turning an experience into a moment of inquiry for the visitor. Let's look at how this can be achieved in spaces indoors and outdoors.

Blinded Participants: How Mazes Work

People rely on visual cues to orient themselves in an unfamiliar environment, but visual cues can be overpowered by other elements of experience. Because wayfinding depends on vision, the absence of visual exposure to an environment can cause people to get lost or to misrepresent their travel when asked to describe their movement through a space. For instance, being able to see something in

the distance doesn't mean that you know the exact route to take to get there. But the landmark provides a visual cue to move toward. Sometimes our participant experience overrides our spectator experience.

In an ethnographic interview I conducted with American tourists (Dewey 2012), I found a situation where tourists were confused about the directions between two points along a straight line because even though they could see down the straight line (a street too long to walk down) toward the destination, when they took an underground train to that destination, the path the train took was angled and not straight. When the subjects were asked for walking directions between the two points, instead of drawing a straight line, they drew an angled line, as if the path had a turn in it. Even though they had clear visual line of sight down the straight street, their experience tricked them. Because they hadn't walked down the street, even though as spectators they had seen that the path on the street was straight, when asked to draw a map between the two end points, their participant experience overrode their spectator knowledge. As participants, they were blinded, and this is exactly how mazes work.

Similar effects can be designed into an experience by creating mazes, tunnels, and other hidden paths through a space if they are built in such a way that they prevent audience members from figuring out where they are in their journey through the space.

Bjarke Ingels's architecture firm BIG installed a large maze (Image 5) in the National Building Museum during the summer of 2014, titled *BIG MAZE*. The design of the maze created the situation that once you got to the middle of the maze you would be able to see the path out of the maze. While you might get lost in the maze because you only have an immersive participant viewpoint for most of the experience, you could eventually find your way out when you got to the center because it provided you with a blended participant-spectator perspective. This was achieved by structuring the walls of the maze so that they became increasingly shorter as par-

Image 5. *BIG MAZE*. Courtesy of BIG – Bjarke Ingels Group. Photo by Kevin Allen.

ticipants walked toward the center. By excavating a "cone of vision" (Image 3) viewpoint was opened up and viewers could see the path they took to the center (Image 4)

Providing Vantage Points to Give Spatial Awareness

People often find it difficult to translate their horizontal experience of a place into vertical map-like knowledge. It's easy to have participant viewpoint because you simply have to occupy the space, but acquiring spectator viewpoint either takes a special kind of thinking which can be provided in the form of a map or, even better, by offering a vantage point that enables visitors to use their bodily experience to acquire new knowledge about a place.

Imagine walking along a path through a park and seeing gentle hills over to the side of the path like those in Image 6. They look like nice hills, but not much more. The view of the hills is from a horizontal participant perspective, and thinking about the way the hills connect to the broader landscape requires you to see the hills from above.

When the path leads to a metal tower with steps (Image 7), climbing that tower affords a view down onto the hills and lets you see how they are connected together (Image 8). In this case, the hills are from burial mounds at the Great Serpent Mound and climbing the tower lets you see the serpentine meander of the hills that is not visible from a participant viewpoint on the ground. Providing a vantage point helps visitors to make sense of the landscape.

Using Viewpoint to Create Feelings and Emotional Responses

In the same way that disorientation and orientation can be built into an environment such as a maze inside of a gallery, they can also be built into larger scale landscapes to create a sense of remoteness or isolation. Dewey (2014) outlined a method for creating artificial remoteness by using path-based approaches that struc-

Image 6. Hillside. RyanDewey.org.

ture the experience by participant viewpoint, spectator viewpoint, or a sequence of both participant and spectator views.

Sometimes a first-person participant viewpoint will evoke a feeling of lostness and remoteness when the landscape closes in around you and seems to swallow you by restricting what and how far you can see in the landscape. The reason immersive participant views evoke the feeling of remoteness is because the landscape prevents you from seeing things as a spectator and you lose the ability to see beyond your immediate location, which strips away your ability to find a reference point to track your movement. Participant viewpoint creates remoteness by hiding everything that is outside of your immediate environment, it is disorienting and works by confusing you or obfuscating the context of your location.

Spectator viewpoint can also evoke remoteness by revealing reference points in the outside world. This serves to orient you because it allows you to see how far you are from some known reference point. As long as the apparent distance and scale seem correct, you can design remoteness and isolation into an environment.

Think about it this way: As you walk into a deep forest valley you start to lose your ability to see where you are because the valley swallows you. But as you climb out of the valley up to the peak of the mountain you gain a clear view that lets you see where you are. Sometimes seeing where you are creates a feeling of remoteness, and sometimes not being able to see where you are creates a feeling of remoteness.

A combination of spectator and participant viewpoints also bring out the sensation of remoteness if they happen with the right sequencing. It may be easier to create artificial remoteness if the sequence of viewpoints reveals and hides the right information at the right time.

One way to design remoteness is to gradually remove the views that provide the orienting information, to slowly disorient by hiding the orienting features. Take a look at Figure 10, which shows a path and container approach to remoteness, to see how this might look in plan view.

In this figure, the path starts on a higher elevation and moves toward a hillside that slopes downward. The circular patches in the diagram are zones that all have different types of plants and trees growing in them. The different plantings provide different views because some plants are taller than others so some planting zones (like thick shrubbery) block out views while others (like grassy lawns) let you see a clear view. In this figure, the large concentric circles are plantings of trees in different densities.

Your initial path is clear and you can see a valley up ahead and you can see a forest that you are about to walk into. These are spectator views that you will quickly lose. You enter the outskirts of a sparsely planted forest which slowly removes your spectator view and begins to immerse you with constrained participant viewpoint as the forest begins to swallow you.

Then your path winds around toward the side of the dense forest and enters the dense forest which completely removes your specta-

Image 7. Observation tower at Great Serpent Mound. RyanDewey.org.
Image 8. Spectator view of Great Serpent Mound. RyanDewey.org.

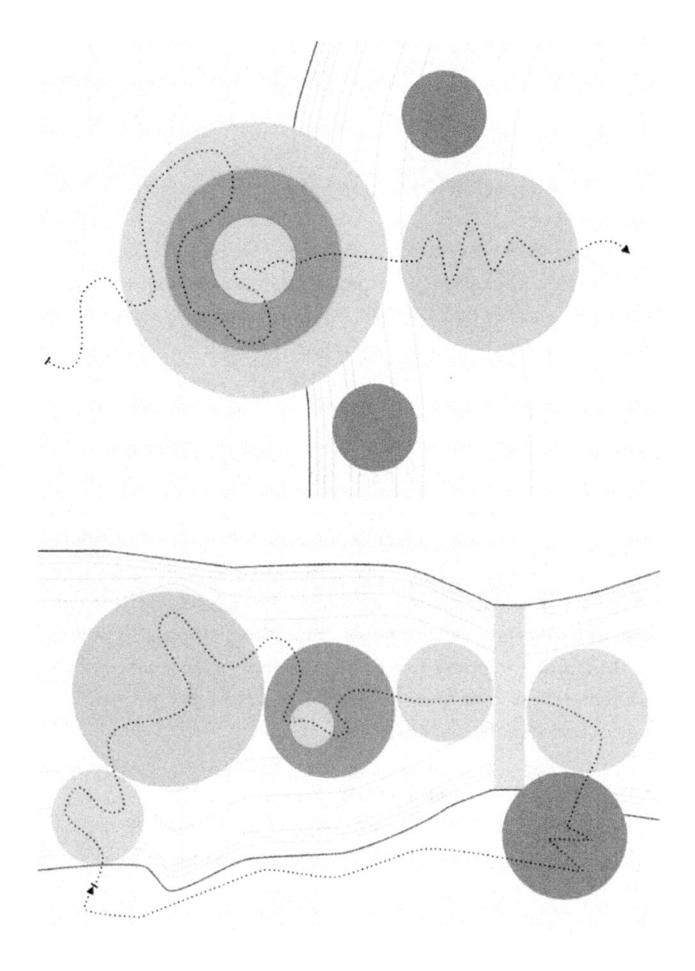

tor view and winds back around toward the other side of the circle before it enters another sparse forest that is like an opening. This opening is completely surrounded by dense forest and it feels like you are farther than you actually are.

The meandering path through the different forests disorients you as your ability to orient with spectator viewpoint is taken away by the trees and the winding path itself. The path curves enough to block you from having a clear view of what lies ahead so you can only see your immediate environment as a participant.

At the moment that you enter the inner circle you feel as though you've walked a long way even though you are not very far from where you began. The decreasing oscillation between spectator and participant views disorients. The experience of being in the thick woods and then walking into the inner circle is opening and expansive, it provides a moment of refuge because the opening is small enough that you get the sense of being inside of a container.

Now it's time to reorient by introducing spectator viewpoint again, giving you the ability to identify a reference point and track your movement with respect to that reference point. In the inner circle your path begins to straighten out as you approach the downward sloping hillside. Since you are at the top of the hill and the forest slopes down with the hillside you can see the swale of the landscape as a spectator and you can also see that your path is a straight shot as it enters a patch of low-lying rushes. Gradually you encounter views that re-orient, or at least they are views that let you see your context in the landscape.

When a path opens up and you can see the view ahead it facilitates a kind of mental simulation (compare with Tool #2) which activates the landscape and gives you the perspective that the landscape has agency. This is often noted in the language people use to describe spectator views of a path-like shape (Dewey 2012), where they will use "fictive motion" descriptions with active verbs like "*the path zigzags across the field*" or "*the path goes this way*" or "*that trail climbs the hillside,*" when in fact the path isn't doing anything active at all other than acting as a support for actual motion (as described in Matlock's *Fictive Motion as Cognitive Simulation*) by a person or an animal or a vehicle. The language descriptions we use suggest that the path is animated and in motion because we simulate that motion. Oakley (2009) argues that these fictive motion descriptions are actually our focus of attention moving along the path shape that we see. The environments that afford spectator views come alive with dynamic language because they engage us differently than the environments that only afford participant views.

Figure 10. Nested Containers Model of Path Design. RyanDewey.org.
Figure 11. Sequenced Containers Model of Path Design. RyanDewey.org.

The nested container model (Figure 10) provides one arrangement that can move people from spectator orientation to participant immersive disorientation and back out to spectator orientation again. Another arrangement would be the sequential container model (Figure 11) that resembles beads of a necklace, where the path is the string and the different zones are the beads.

The path starts on the side of a steep slope down into a valley and from the beginning of the path, you are able to see the lay of the land and the different zones that you will be walking through. This path is a closed loop and it will return you to where you began. You can see the bridge in the distance and as you start to climb down the hillside, you lose sight of the spectator view and become immersed in the participant view. By the time you reach the boardwalk in the large zone, the tall reeds and the tops of the trees ahead block your view of the bridge and you lose that as a reference point. Just before you enter the woods, the path swings out to allow you to see the bridge again briefly before plunging into the forest. You walk through the forest and come upon an enclosed opening, a meadow in the middle of the woods. Back on the trail, you walk out of the woods into an expansive view of the towering bridge. The contrast between being in the thick woods and walking out into the open grassy area only heightens the experience of the bridge being focal in your field of view as the trail takes you directly toward and under the bridge. As you start the walk back up the hillside, you cross through a gravelly slope that gives you clear view of your path ahead.

The distinctive element of this model is that the path back to the trailhead takes you past the different zones that you walked through in the valley, giving you a spectator view that enables you to make sense out of where you had just walked. From the top of the hillside, path shape is visible in many of the zones. This sequential container model begins and ends with clear spectator views that help make sense out of the experience, and then the path takes those third-person spectator views away and immerses you in participant first-person views, only briefly providing glimpses of your context from a spectator vantage point, and then pushing you back into participation.

These models are very much like a maze, where you lose the ability to see how your position at any point in time is related to your position at another point in time. It is disorienting, but only to you. If someone is watching you move through the maze or the environment, their spectator viewpoint helps them see how your movement and your position are connected. The oscillation between spectator and participant view creates a dynamic situation for you as the person experiencing it and this shows up in the type of language descriptions people use to describe the features of the landscape ("the path twisted back...," etc.). Moving people back and forth between viewpoints is an easy way to evoke orientation, disorientation, and re-orientation. If sequenced appropriately with the right kinds of supporting environments, these three states can provide the structural framework to create paths that evoke feelings of refuge, remoteness, awe, isolation, confusion, tranquility, peace, ease, tension, uncertainty, and even freedom.

Multi-Sensory Participant-Spectator Viewpoint

Participant and spectator are not merely visual viewpoints: they are available in other sensory domains like sound and touch, even smell and taste. The visual domain for participants is a sort of visual immersion in a scene, whereas the spectator is outside of a scene, observing the scene as if it were an image. You can think of the visual domain as a container of sorts, where participant viewpoint is inside the container and spectator viewpoint is outside the container. This inside/outside container schema holds for the way we experience participant viewpoint and spectator viewpoint in all of the other sensory systems.

Mix-and-Match Sensory Viewpoints

Combining viewpoints in one sensory system with viewpoints in another sensory system can create interesting effects, heightening and diminishing and confusing the senses. For instance, a visual

	spectator	participant
Visual	observing scene from above, at a distance, mediated through lens or screen	being in scene (immersion), taking first person perspective of an environment
Auditory	hearing noises not caused by self but which occur naturally in the environment (diegetic), or hearing noises not caused by self but which do not occur in the environment (remote listening as quasi-non-diegetic sound)	hearing noise you cause or which emerges naturally from your immediate environment (diegetic), or non-diegetic through use of headphones
Tactile	visually touching a texture, imagining how something feels, describing an observed reaction to texture (e.g., slow and difficult manner of motion over or through a sticky substance)	touching with your own body (active) being touched (passive)
Olfactory	smells wafting from elsewhere, distant smells	smelling your immediate environment or smells you create
Taste	tasting aerosolized particles while being outside of the space where they originate from empathizing with someone eating something you know the taste of (9v battery, lemons, tin foil, soap, etc)	tasting what you put in your own mouth tasting aerosolized particles while being in the space where they originate from
Sensory Vitality	All sensory stimuli move from moments of salience to becoming background sensations regardless of spectator or participant viewpoint. One way to keep a salient sensory stimuli salient longer is to move back and forth between spectator and participant viewpoints of the same salient sensory stimuli.	

participant with a sonic spectator viewpoint would experience a mismatch in sensory feedback for their physical actions. The sounds that are made by the actions their hands are doing (finger noises, tool noises, etc.) might sound to them as if they are coming from further away or from a different location.

The rubber hand illusion illustrates a crossing of visual and haptic sensory stimuli that might have relation to spectator and participant viewpoint. In the illusion, a person puts one hand under a table and above the table a rubber hand is placed in the same configuration as their hand that is under the table. The administrator of the experiment strokes the hand under the table with a feather or a brush while simultaneously stroking the hand on top of the table with the same kind of tool. Eventually the person will begin to feel as if the rubber hand is their hand (they believe it to be the hand that is connected to their body). The hand that they are looking at (spectating hand) becomes the hand that they experience as their hand receiving the sensory stimulation (participant hand). The mind can be tricked by combining a spectator viewpoint in one domain (in this case, visual) and a participant viewpoint in another domain (in this case haptic). The experiment usually ends dramatically with the administrator picking up a hammer and whacking the rubber hand to the subject's surprise.

The Sensory Viewpoints

Table 3 provides a rough, non-exhaustive list of possible participant-spectator distinctions across the senses (spectator viewpoint is presented first, participant second).

It is easy to imagine how crossing these spectator-participant sensory experiences can produce both confusing experiences and experiences with heightened clarity. It is also possible to substitute one sensory viewpoint with another sensory channel in the same viewpoint—to translate one experience into another experience. This happens in every cooking show on television.

Table 3. Sensory-Based Spectator-Participant Structure.

When you watch a cooking show, you experience most of the content of the show as a spectator. Visually, you are watching a scene from afar, but then the camera pushes in to a close up and you can see the technique the chef is performing from a first-person participant viewpoint. For the touch, smell, and taste, you are a spectator because television, film, and screens prevent you from having these sensory experiences. This limitation of the technology is something that the chef makes up for in a performative way, translating for the viewing audience the sensory experiences that you would have if it were you doing the cooking. As a viewer, your spectating role enables you to experience something which you cannot experience as a participant because the chef is performing those sensory experiences for you. The chef might describe the way the texture of a mixture changes (something you could tell by touch and proprioceptive experience through a mixing spoon) by saying something like, "it's starting to get thicker," and perhaps letting you see the effort being put into the process. Or, for smell, something like, "I'm starting to smell the sausage, it must be done," or, "it really smells earthy, like that fresh smell in the forest just after it rains."

In order to represent taste, the chef must make sounds, gestures, and reactions that translate for the viewer the moment of delight and disgust the chef is experiencing. The chef's response to tasting food enables you to see what it tastes like. Then the chef describes the flavors of the food in literal descriptions and in metaphors and these descriptions help you imagine the taste which prompts you to want to make the food yourself.

Methods for Augmenting Viewpoint in the Experience

Changing a person's viewpoint changes their experience. Because viewpoint is a continuum, and it shifts dynamically throughout experience based on physical location and individual attention patterns (e.g., leaning in to get a better view), you can alter a person's viewpoint to change their experience. For instance, you could give a participant viewpoint to someone who is a spectator to make them feel immersed in the environment, and this is what first-person film does: it blends your spectator/viewing reality with the participant/immersed reality. Other combinations of viewpoint switching can be divided into combinations of views for individuals and combinations of views between individuals.

You can create combinations of views *for* individuals by:

— switching a participant to a spectator;
— switching a spectator to a participant;
— taking away all viewpoints (no participant or spectator);
— giving both viewpoints (both participant and spectator); and/or
— taking away one viewpoint from an individual who had both.

You can also create combinations of views *between* individuals by:

— giving one spectator the viewpoint of another participant;
— giving one particular participant the viewpoint of another particular spectator;
— switching participant views between different participants;
— switching spectator views between different spectators;
— switching views between different participants and spectators in an ad hoc manner; and/or
— switching views between different participants and spectators in a controlled manner.

Note: There are numerous examples of artists building these types of interactions. Madeline Schwartzman's *Seeing Yourself Sensing: Redefining Human Perception* (2011) presents a good collection of examples of how artists have operationalized some of these ideas, as does Caroline A. Jones's *Sensorium: Embodied Experience, Technology, and Contemporary Art* (2006).

Remember that spectator/participant viewpoint is not just visual, but is also a distinction that shows up in all of the senses, and you can switch viewpoint for any or all of the senses.

As an example of how you could switch viewpoints between different members of the audience (whether participant or spectator), consider changing what people hear. Use headphones or earpieces to give the audience access to sounds that someone else is hearing. Arrange it so that no one hears the sounds of their local environment, but that they hear the sounds from the perspective of someone else in the experience, so the locus of hearing is shifted out of the personal body and is distributed to someone else's body and their environmental location. The effect gives everyone in your experience a sense of eavesdropping.

Examples of some of the other combinations include: switching visual views by use of cameras and monitors or wearable computing devices, or by taking away viewpoints by creating isolation in one sensory domain like vision (place the audience members in a small room or cover their eyes, but keep them in the environment of the experience). You could even switch the audience member's viewpoint from participant to spectator by changing their location to be outside of the center of action, or you could switch viewpoint in the opposite direction: from spectator to participant, by placing them (physically or by telepresence) inside the center of participant action, and giving both viewpoints to a participant by providing a paper map that shows spectator viewpoint.

Summary

Everything can be divided into two viewpoints: participant viewpoint, which is an *immersive* first-person perspective, and spectator viewpoint, which is a *removed* viewpoint. In path design through an installation, a blended viewpoint of participant and spectator allows visitors to see where they are in the installation, but if available viewpoints oscillate back and forth from participant to spectator to blended and so on, the effects can range from orienting to disorienting, even creating a sense of artificial isolation or remoteness. When viewpoints change throughout an experience, the result is an experience of extreme flow. Controlling flows of viewpoint gives you, as the author of the experience, another angle from which to create a multi-faceted experience. Playing with viewpoint does not need complicated technology or even high budgets. Simple interventions using photographs or even paper maps are more than sufficient to serve the story you are telling. Simple variation in viewpoint throughout an installation will set up your audience to experience rhetorical effects of orientation and disorientation. Blending viewpoints throughout an experience in certain sequences can even evoke moments of cognitive simulation, which may heighten memory of the experience and the imageability of your installation. Further, viewpoint can draw on multiple sensory systems and even be used to create moments where one sensory system provides a viewpoint that conflicts with another sensory system, creating complex series of matches and mismatches to serve the rhetorical effects that the designer has in mind.

TOOL #4
Embodiment

What is a Body?

Right now, without looking around, try to feel all of the parts of your body in succession. First, feel what the tips of your fingers are feeling. Move up through the fingers to the palms of your hands. Can you feel the tops of your hands? Is there a temperature difference between the bottoms of your hands and the tops of your hands? What do your wrists feel like? Move up the arm to the elbow and notice the balance of weight between the forearm and the upper muscles and shoulder. Imagine the connection at the shoulder and feel the space between your shoulders, and then move back and forth between the shoulders. Next, imagine your spine shooting downwards toward the floor: can you feel your ribs curling around from the spine toward the chest? Breathe. Does your breath make you more or less aware of your spine? Does it make you more or less aware of your chest? Without looking at your stomach, do you feel it? Is it hungry? Is it satisfied? Are you thirsty? Do you know the difference between hunger and thirst, or do you mistake one for the other? Is your bladder full? How does that affect your ability to concentrate, or your ability to be patient with people? Your mood? How aware are you of the flesh of your actual genitals on a moment-by-moment basis? Do you feel your hips right now? Imagine the mechanical connection of your hips and feel your legs attach to the torso. If you are standing, where do you feel the weight? If you are sitting, where do you feel the tension? Is your body being pulled forward, curling in on itself, or is the attitude of your body upright and open? Are your thighs tight? Do your knees open to the sides? Are your lower legs twisted or straight? Is there tension in your ankles? Can you feel your feet? Is this your body? Lift your head upwards and look at the ceiling or the sky. Think about the tightness in the front of your neck as an index of a life spent looking at objects like screens and papers and tasks that you watch yourself doing rather than a life spent looking around. You need to look up more. You need to look around more. Do you enjoy your body? Do you want to? How does your body make you feel about your life? How often are you aware of the physical structure of your body? Is pain the only thing that awakens you to realize your body is there? Do you also recognize pleasant sensations? Do you also recognize no sensations? Do you need your body to move? Do you need your body to sense things in the world? Do you need your body to think?

Embodied cognitive science argues that the body is part of the mind and that we use our bodily experience of the world as a way to think about the world. The body is not just connected to the mind, because the body *is* the mind. We think by acting. This is helpful for experience hacking because it opens the possibility that, to some extent, although with exceptions, creating a particular experience that engages the body can result in people thinking about or interpreting that experience in a somewhat predictable way (as seen in Sato et al. 2013, and in Bergen and Wheeler's 2010 study on toward/away motion). Action is a backdoor to cognition. Action is a portal to hacking experience. Let's explore the reasoning behind this idea by first looking at the way the body shows up in the language we use to describe the world.

In the mid-1980s, cognitive linguists began analyzing the language people use to describe abstract concepts and they discovered that a lot of abstract reasoning relies on concrete examples from real-life, everyday experience. They started noticing a systematic language structure between abstract concepts and concrete concepts that are now called *conceptual metaphors.* The principal argument from this theory is that we use our embodied experience as the basis for our emotional, abstract, and philosophical experience. Since then, linguists have been systematically categorizing the types of conceptual metaphors people use and have been looking at the image-schematic structure of those metaphors for insight into the ways that people use their body to think about less physical ideas. The evidence from these studies points to the body playing a primary role in the way we think about our experiences.

What Do You Mean by Embodiment?

Since the body and the mind are unified, we use our bodies to make meaning as we experience the world around us. We understand what experiences mean for us by using a blend of our sensorimotor skills (or the way we use our senses and our bodies to engage the world) and perceptual attention. The body is the sense organ *and* the sense-making organ. Our mind is part of our body. There is no separation between the mind and the body. Embodiment is the unification of the mind with a body, in a body. The mind is embodied in a corporeal form (i.e., a body). That body-mind combination engages in thinking by engaging in acting, doing, performing, moving, behaving, and so on.

Research on conceptual metaphors shows that our conceptual system is linked and shaped by perceptual and motor systems and the only way people can form and frame concepts is through their bodies. This means that sensorimotor experience influences our thinking or reasoning about the world. One conclusion from this idea is that diversity in physical structure leads to a form of cognitive diversity. Different bodies result in different ways of thinking about the world. Another conclusion is that our knowledge and notions of truth depend on our bodies (Lakoff and Johnson 1999, esp. 555–57 for complete discussion, and 551–54 for a comparison with the obsolete disembodied view of the mind in all forms of mind-body dualism).

Access Your Audiences' Abstract Reasoning by Altering Their Bodily Experiences

We think with our bodies. Our embodiment has been thoroughly demonstrated to shape our ability to reason about the abstract elements of the world (Lakoff and Johnson 1999). These reasoning patterns pervade everyday thought and experience. It makes sense, then, that hacking into these patterns can give you access to shaping an audience's thought processes to some extent. This access ultimately helps you evoke responses that are meaningful to people in your engineered experience. It is possible to create new patterns of reasoning by combining sensory, emotional, and physical elements and these will be seen to have powerful effects on personal experience. Throughout the remainder of this book, methods will be presented that build upon concepts already introduced to teach you to build new reasoning structures and metaphors between concrete embodied experience and abstract experience for the participants in your engineered experience.

The language that people use to describe the world reflects how we use the concrete world to think and talk about the abstract world. We use concrete experiences (for example, a journey) to talk about less tangible experiences (like, life in general, maybe our relationships, or internal states) when we use descriptions like: *his life took a turn for the worse, they're on a crash course now, she's trying to find her way, or I'm feeling lost.* When we say those things, we use the experiential world of **travel** to make sense out of the abstract world of **life**. If you can get people to think about a complex idea or abstract quality in terms of some concrete experience, then you establish a language (verbal, visual, or otherwise) that has some bodily basis in reality. Following the language of this abstract-concrete relationship as a framework for experience design, you can build the experience according to the inherent logic of the relationship and know that it ties into human cognition to help people interpret the abstract.

Language often reflects experience—for example, with a painting of some visual scene that features long linear leading lines, it would be expected that people would describe the scene by using fictive motion descriptions because of the cognitive simulation that takes place in the experience of the work of art. In fact, this holds up in the research that shows that cognitive simulation includes simulated motion along trajectory paths in visual scenes (Talmy 2001; Matlock 2004b) and in images of physical space (Oakley 2009). Formal salience in image composition evokes cognitive simulation because of embodiment and embodied knowledge. Tying this together with Noë's (2004) argument about the role of motion as a means to acquire content (or meaning) about a physical scene, it is reasonable to say that the formal salience of the physical composition of an immersive installation should also tap into the dynamism of things like fictive motion experiences in physical spaces (and not just in images). As we've already seen, initial results in Dewey (2012) show that this might be the case in outdoor path-based environments as the composition of the scene oscillates back and forth between providing spectator and participant viewpoints. **This means that compositional elements of physical space are inherently tied to embodiment, and that physical spaces which better organize those elements will have stronger rhetorical value on the people experiencing those spaces.**

Using Embodiment to Tap into the Mind

In order to engineer experiences that affect people's emotions and state of mind, we need to be able to get into people's minds. If language and conceptual metaphor are interfaces for our cognitive states, perhaps we can crawl through these open windows of language and conceptual metaphor and activate or change the concrete world in places where the concrete world has been seen to shape the abstract world.

If something concrete is used metaphorically to describe our abstract experiences (such as emotions and love or indifference), then we can shape the concrete environment to shape the way people think about (and talk about) those abstract ideas. If we shape the environment successfully, then we can change the way people think about the world and we will have hacked experience. Metaphors are one of the primary tools for this because they connect to basic embodied experience through image schemas. We will see in Tool #6 how sensory conditions of our environment can influence the abstract reasoning patterns people use in the world. Shaping the environment means more than physically altering the environment, it also includes shaping the way that people approach the environment through priming activities and framing audience expectations. You can give people a perspective on an environment by using conceptual metaphors that shift the way people see the environment toward the way you want them to see it. This is the role of the metaphor designer. Museums do this all the time with careful wording on labels that incorporate the types of lenses the museum wants you to use when viewing the work. The language in the labels directs viewer attention to aspects of the work. They tell you what to pay attention to and how to think about it. Label language shapes your experience of the artwork you are viewing. Since language can direct attention, hacking into experience can be as simple as using language that elicits the responses we want to achieve. Pay attention to the ways that content, structure, and delivery influence the response and adjust accordingly.

Language acts as a window into the mind because it reflects conceptual structure (Evans 2009). Because language reflects cognition, it might be possible to change cognition by changing the way language is used. This idea is somewhat controversial depending on your assumptions of how cognition operates as a system. What is less controversial is that using language entails framing information for communication and the result of the communication can bring about change in the real world. It is hard to argue with repeatedly validated empirical research which shows how structures in language directly impact cognitive skills. In this case, metaphor research has demonstrated that people routinely use the concrete world to reason about the abstract world (Lakoff and Johnson 1999; Slepian and Ambady 2014; Lee and Schwartz 2012).

In our journey toward designing experiences that evoke responses, exploring the basic structure of metaphors will help show how experiences can be enhanced by accessing conceptual and perceptual structure in the audience through design elements in the physical location of the experience.

We Use Our Bodies to Think

One tool that helps us understand the role of the body in cognition is the type of language that we use to describe our experience of the world. Using language as a tool to look at cognition helps illuminate how we reason about abstract ideas like love, curiosity, good and bad, meaningfulness, categorization, time, and many others. We tend to use concrete objects and systems to think about abstract notions and ideas, which is convenient for artists as we anchor concepts in the material objects and experiential worlds we create. It makes sense for us to use the physical to explicate the abstract.

Research in cognition has uncovered a series of language structures that tie directly to our body and embodied experience of the world. This link between embodiment and language shows up in the conceptual metaphors we use to talk about abstract concepts. Lakoff and Johnson (1980, 1999) uncovered this link with their work on primary and complex metaphors. They argue that very basic physical skills learned as children through activities like stacking blocks, putting things away, walking from one point to another, and touching things form the basis for thinking about abstract ideas like *good, happiness, importance, categories, similarity, purposes, relationships, knowledge,* etc. This connection between physical skills and abstract ideas is seen in some of the following primary metaphors (a more extensive list can be found in Lakoff and Johnson 1999, 50–54): *More is Up, Categories are Containers, Similarity is Closeness, Linear Scales are Paths, Organization is Physical Structure, States are Locations, Time is Motion, Change is Motion, Purposes are Destinations, Causes are Physical Forces, Relationships are Enclosures, Control is Up, Knowing is Seeing, Seeing is Touching, Understanding is Grasping.*

These primary metaphors show up in systematic ways in our everyday language and we reason about the abstract world by using these metaphors in our casual speech. All of these metaphors directly trace back to some basic experience we have had with our bodies in the world. They even show up outside of language in behavior, motion, design, and objects (Lakoff and Johnson 1999, 57). Part of what makes these metaphors so ubiquitous in everyday life is that they come out of primary experience and are structured by image schemas (as mentioned elsewhere in this chapter).

Consider these examples of the primary metaphor system *Good is Up* being used in language:

Good is Up (entails Bad is Down)

— **emotional states**: his mood was elevated, she lifted your spirits, she is depressing, you can't bring me down, he's in the dumps, I fell into addiction but now I'm climbing out of it, you seem like you're on the up and up (etc.).
— **evaluative judgments**: that was low, high brow/low brow, underhanded, aboveboard, under the table, under the counter, etc.
— **trajectory paths**: he's crashing, it's a downward spiral, that's a slippery slope, you're just trying to pull me down, I'm flying high, I'm taking off, you can't bring me down, ascending/descending, rising/falling, etc.
— **hierarchical structure**: climbing the ladder, high on the food chain, superior/inferior, underlings, top-shelf, bottom rung.

Because conceptual metaphors have an image schematic basis and are grounded in our bodily experience of the world, it is easy to make connections between the language metaphor and the same structure in multi-sensory metaphors. Multi-sensory metaphors need to take the image schema as an armature for the experience. For instance, light and sound are two sensory channels that we often think of in terms of up and down, and you can see this in the way we talk about light and sound:

- **lights**: bring up the house lights, turn the lights down low, dim the lights, the sun is rising, the sun is setting, etc.
- **sound**: raise the volume, turn the volume down, low notes, high notes, mid-range, climb the scale, ascending/descending, crescendo, pump up the volume, raise the tempo, feel the bass down low, etc.

Now think about the ways that light and sound can be paired together at an event like a concert. Stage lighting and house lighting often increase in brightness in sync with volume and tempo in the music. Light and volume also often peak at the same moment. That upward journey toward peak is a dramatic path that is both emotive and exhilarating, sound and light build off of each other to convey the experiential metaphor of *Good is Up*. When they peak, it happens to coordinate with an important moment of the song and this is where form meets function in a multi-sensory metaphor. Our bodies understand this metaphor because we experience things that are up (or bright, or loud) as good in our physical and conceptual world (note: this is not to say that there isn't also a parallel metaphor where *Good is Down,* just that there is a systematic set of relationships that exist for *Good is Up*). Context helps determine which metaphor is active at some particular moment. (More on this multi-sensory pairing of conceptual metaphors to sensory stimuli in Tool #6.)

Our bodies make connections for our conceptual world. Conceptual metaphors are rooted firmly in our bodily experience, to our primary experience of the world. The physical gives birth to the conceptual. Because of this link to primary experience, it might be possible to dismantle the metaphor, to use it as a backdoor to primary experience, and then use it as a tool for building new experiences. In fact, not only do we map the abstract world onto the concrete world of physical experience, but we can also reverse it and map the physical onto the abstract (Slepian and Ambady 2014). In other words, we can use the concrete world to think about the abstract world, and we can use the abstract world to think about the concrete world.

Not all cognitive linguists agree that conceptual metaphors are bi-directional, but experiments have begun to show that it is possible. Slepian and Ambady conducted an experiment in which people experienced a controlled effect in their embodied experience and were taught a new metaphor that corresponded to that controlled experience. It was a metaphor that they could not learn from bodily experience because it has no grounding in our natural physical world. They created sets of opposing metaphors like "the present is heavy" vs. "the past is heavy," which suggests that units of time (e.g., the present, the past) are objects that have weight (e.g., are heavy). Obviously this is not something that we experience in everyday life—we don't connect time and weight, but subjects in this experiment were able to learn this metaphor and use it to think about the world. The subjects each read a few sentences which conditioned them to either think of the past as having weight (e.g., "You must carry your past with you wherever you go") or the future as having weight (e.g., "The decisions of your past carry no weight. It is your decisions today that define who you are..."). Subjects in each group then had to estimate the weight of a book. The idea was that the subjects who were conditioned to think of the past as heavy and the present as light would judge an old book to be heavier than the subjects who were conditioned to think of the present as heavy and the past as light, which is what happened; subjects learned a novel metaphor system, which influenced their estimates of how much a physical object weighed.

This means that metaphors that the subjects learned lead to what is called **embodied simulation**, or, as Slepian and Ambady put it, "by learning an embodied metaphor, sensorimotor states become associated with the abstract concept" (2014, 4). In other words, ***some metaphors can create sensations***. Other evidence points to this; for instance, Lee and Schwartz (2012) demonstrated that *suspicion* (an abstract concept) could be induced by presenting people with *fishy smells* (a physical sensation), and also that ability to detect fishy smells increased when people were primed to be suspicious. By feeding the abstract concept to people, they were better at identifying a concrete sensation. This shows a clear bi-directionality of the metaphoric expression of suspicion in "something smells fishy." By controlling peoples' olfactory experience, they were able to evoke a particular emotional experience, and by controlling emotional appraisal (by priming people to be suspicious), they were able to evoke a heightened state of sensory perception (a heightened ability to identify smells).

These experiments suggest that linkages between the physical world and the abstract world can be created in people and their experience of the world. Sensations and emotions can be induced by physical elements, and they can be induced across sensory channels (they are multimodal). While these findings need continued research to understand the full scope of implications, current research suggests a fortunate opening for the use of metaphor as a means of evoking perceptual and emotional responses in people. What we know at this point is that Slepian and Ambady's model tentatively suggests that repeated exposure to a new metaphor can increase the effects of the metaphor (2014, 9), and that perhaps it is the strong image-schematic structure of metaphors that enables people to map the abstract into their embodied experience (2014, 9).

Metaphors provide a window into the mind through the body, enabling environments and experiences to have emotional and mental influence on your audience. These metaphors will create links between the concrete and the abstract in order to map different sensory systems and experience for controlling ambient attention and emotional tone.

Before moving on, it is worth mentioning that because conceptual metaphors are structured by image schemas, they frequently show up in 2D forms, especially in forms that have strong compositional structure. Elements of composition in artistic works will map to conceptual metaphors by way of image schemas. For instance, Lakoff (2006, 155–56) translates Arnheim's description of Rembrandt's *Christ at Emmaus* in terms of conceptual metaphors, noting that the image of Christ and the images of the other people in the painting exhibit nesting relationships of containment in two triangular compositional arrangements. The metaphors *Important is Central, Divine is Up, Humility is Down, Morality is Light,* and *Knowledge is Light* structure the visual content and ascribe conceptual content to the painting. Lakoff explains these metaphors as emerging from orientational image schemas: *High-Low, Container* (x2), *Center Periphery* (x2), and *Light-Dark* (2006, 156). The most interesting analysis is of the social relationship of Christ to the servant giving food to Christ. Christ is being served by a boy who is socially below Christ but positioned above Christ's head, accessing the visual metaphor *Humility is Down,* ostensibly giving importance to the humble act of serving Christ while simultaneously recognizing the humility of Christ. In this way, Rembrandt has focused on Christ's humility through spatial composition.

Because image schemas and conceptual metaphors are recruited into visual works, viewers have immediate access to notions of embodiment by engaging those works. And since conceptual metaphors originate in embodied experience, perhaps a visual metaphor will create sensorimotor sensations through embodied simulation, reifying the concept through bodily perception and a possible link to empathetic mirroring.

Apply This Now

How you deploy metaphors is up to you. Conceptual metaphors can be deployed in immersive environments (such as linking a fishy smell to a scene that reeks with suspicion), or they can be used in static visual 2D imagery (such as translating the image schematic structure into elements of visual composition). Remember that a path experience is a sequence of moments and each moment can draw on principles from 2-dimensional composition. Perhaps you can find ways to bring 2D composition into the fabric of an immersive environment, since immersive environments make use of 2D forms.

Get started thinking about the ways metaphors work for your subject. Perhaps start with an abstract concept that you want people to experience (e.g., affection) and look for things in the physical world that can anchor that concept.

Regarding the conceptual metaphor *affection is warmth,* Lakoff and Johnson (1999) argue that the sensorimotor domain of *temperature* is something we learn to associate with the abstract notion of *affection* through our embodied experience of "feeling warm while being held affectionately," presumably something learned early on as an infant is snuggled and swaddled and held against the mother's chest. Take this idea of proximity and heat correlating with affection and begin to design an experience with it.

Perhaps you want to evoke affection between people, even strangers. Start listing experiences, objects, and conditions that create warmth (in terms of body temperature or ambient temperature) and determine which ones might prompt people to feel affection. Determine what else you need to evoke that feeling of affection, or perhaps identify other physical elements (beside temperature) that might bring a feeling of warmth and layer them into the experience to reinforce and corroborate the stimuli toward an overwhelming sense of warmth. Create situations of social proximity where people feel that it is safe to let their guard down. Then begin to think about ways to elicit affection. Maybe you give participants a script that they have to follow, maybe you place random participants in enclosures that create warmth through proximity, or maybe you do something simple such as giving each participant a fortune cookie with a fortune that suggests something about another participant. Come at the idea of *affection is warmth* from as many directions as necessary to evoke the affection response in your participants.

You could even use these metaphors as tools with confederates who mingle among the crowd to perform emotional contagion. If you have a script that is based on a conceptual metaphor that is grounded in bodily experience, make the bodily experience explicit in the interaction between your confederates and the audience members. Then support that metaphor system by structuring the visual aspects of the experience in ways that reinforce the metaphor. Also consider using language that makes use of the metaphor system in some way to prime people for experiencing that metaphor.

Embodiment and Knowing the Body: Proprioception as Form of Self-Attention

Another type of attention exists, and it is subtle, but you feel it when you do things with your body. Sometimes when people use the word "biofeedback," they are talking about proprioception, which is an attention system that accounts for how you know what you are sensing with your body, whether it is your awareness of your posture, the feeling of radiated heat on your face when the sun shines in your window, or the awareness of micro-motions when doing delicate work. It is a form of kinesthetic intelligence, and some people have a more fine-tuned proprioception than others.

Without simplifying this too much, consider proprioception to be the signals that you experience inside your body when you experience your body as an object— when, e.g., you pay attention to the fact that your body really exists (we are often aware of our bodies in moments of pain, but movement, motion, and pleasurable sensations also bring our bodies to the front of our attention). Proprioception relates to knowing the limitations and capabilities of your body in the moment, and it helps explain how some people understand their bodies better than other people. Because different people have different sensitivities to the states of their bodies, designing an experience that relies too heavily on proprioceptive attention might decrease the effect of the experience for a wide audience, perhaps even deadening any figure-ground contrast you are trying to establish. But if your goal is to increase proprioceptive knowledge in all of your audience members, then design meditative and reflective experiences that help everyone heighten bodily attention.

Image 9. Institute for New Feeling, Insole. Courtesy of the artists.

The Institute for New Feeling is an art collective that explores the senses and activates proprioceptive and empathetic experiences in viewers. One of their early works included a moleskin insole (Image 9) that people could adhere directly to their foot that would apply selective pressure on different zones of the foot, which correlated to a schematic chart (Image 10). It was described as a therapeutic device.

This insole augments embodiment prosthetically by causing the activity of walking to become a form of therapy, rather than merely ambulation. Movement excites pressure-points on the feet to deliver repetitive therapeutic moments. But this is not just any type of therapy. It is also a branded therapy with strong capitalistic inclinations, noted by the throwaway nature of the insole itself and an admission from the artists: "An adhesive moleskin insole that functions both as an invisible advertisement and a pressure-point therapy. Borrowing elements from reflexology, corporate branding strategies, and Dr. Scholls, this flesh-toned foot pad is applied to the bottom of a participants' feet, allowing our logo to permeate their life in an ever-present, yet therapeutic way."

The type of therapy delivered through this device made direct use of proprioception, but coupled it with a rhetorical goal of getting the user to question how corporations design products that our bodies adopt as part of the proprioceptive system.

Thinking through Action

Modern theories of cognition consider the mind and the body to work hand-in-hand in reasoning about the world. While this sounds intuitive, for the majority of the last several hundred years, the Western world has held tightly to a notion of a disembodied mind which argued for reasoning and meaning that did not depend upon a body. It forms the basis of many approaches to artificial intelligence that operate under the outmoded assumption that a body isn't required for human intelligence.

This embodiment hypothesis—that our bodies work with our brains to provide the structure for cognitive experience—means that variation in bodies within a group of people results in each person having some minor variation in how they experience the world. Our bodies limit our ability to access some parts of reality, and importantly, this includes concepts. Think about a concept like time. You have experience of days and years, even decades and lifetimes, and the older you get, the more familiar you are with how long a year actually feels. So when people talk about something that happened five or ten years ago, it is fairly easy to imagine five or ten years. But when the conversation switches to discussions of thousands, millions, or billions of years, you don't have the experience

Image 10. Diagram of pressure points, Institute for New Feeling. Courtesy of the artists.

necessary to find that scale relatable. This is partly why one of the greatest problems in geoscience education is getting students to be able to understand geologic time. Our embodiment limits our experience on a temporal scale (Semken et al. 2009).

Universality and Diversity of Embodiment

Embodiment shapes experience of the world. As a species, *Homo sapiens* have a common embodiment that is largely stable across the human population. The default body includes two arms with opposable thumbs, two legs that support standing, differentiated fingers and toes, no tail, arms that extend with a reach that measures approximately that of the height of the person, two eyes that see within the same wavelength, ears that hear in a certain frequency, a voice that has a roughly stable range, taste receptors that distinguish between five basic flavors, and a nose that explores the hundreds of thousands of scents that can be distinguished by people. On a gross level, humans are all basically the same. This is the universality of the body: we all have something in common.

There is also a diversity of embodiment that ranges between cultures, professions, maturity levels, and sizes. Not everyone is the same size which leads to a lot of variation in ability. On a crude level, taller people can see farther than shorter people, have better leverage in lifting things and bigger embraces, can't fit into small places, and have wide strides. Even a person's profession can alter embodiment. While certain bodies might predispose a person to particular professions, professions also refine the body through practice to experience the world in refined ways. Consider the differences in the ways that bodies are tuned for different professions: athletes have bodies that are honed to perfect their sport, perfumers have noses that are trained to untangle scents, chefs master flavor and visual composition, and office workers have bodies that tolerate cubical containment, bad ventilation, complex office noise, and fluorescent light.

This is not a trite observation. All of these variations exist in the different members of the audiences you are trying to hack. Some will have bodies that resist or comply more than others with your intentions. If you want everyone to have a similar experience, then your goal will be to offer equalizing experiences that give everyone the same access to the intervention so that the environment controls the experience as much as possible. If you want to highlight diversity of embodiment, your goal will be to offer experiences that include and exclude different bodies so that the body determines the experience. Take these differences into account when creating experiences that augment, mute, or modify embodiment. Perhaps the easiest to control is viewpoint.

If the way we think about the world is partly determined by our bodies, then meeting the body halfway and tailoring experiences to fit the body should result in more intuitive and meaningful experiences. Broadly speaking, humans as a species share similar body structure. There is variation in age-related size (relative height and weight); genetically-determined size (mature height and weight); weight resulting from factors of choice, economics, or politics; bodily proportions; skin color; age; physical strength, etc. Then there are factors of bodily augmentation: height can vary based on type of shoes, skin appearance alters in exposures to sun and chemicals as well as in varying levels of hydration; age can appear to be older or younger based on fashion, diet, makeup, surgery, and even social context (such as three sisters ages 18-30 hanging out and appearing to be of relatively the same age). These points of variation result in differences of experience, and can serve as rhetorical tools in the composition of art, events, and experiences.

Our bodies and our experience of the world shape our cognitive abilities. When we encounter concepts that are beyond our experience, we often need help comprehending those concepts. In this **embodied cognition** hypothesis, it would seem reasonable that if we can't experience something directly and it is beyond the sensory reach of our bodies, then it would be foreign and difficult to understand. This presents opportunities and challenges for the artist. The opportunity exists because choices are wide open when looking for things that are outside the scale of normal experience, but the challenge is finding a way to translate it in terms that people can grasp. You want people to relate to something that is not at human scale as if it were at human scale.

Experiment with embodied cognition to get participants thinking through action by providing activities for the audience that require them to figure out how something works by playing around with it. Provide objects that can be physically manipulated and let them learn something by touching it, or experiencing it with some of their senses, or possibly removing certain sense systems. Change the experience into one that requires them to fill in the missing senses by compensating for one sense with another sense.

Manipulate Viewpoint

One immediate way of changing embodiment is by giving a spectator a participant view, or by giving a participant a spectator view (as in a drone operator's viewpoint, by way of a kind of telepresence), or perhaps removing one viewpoint (as would happen if a GPS ceased functioning mid-course in a chaotic city, or how pilots fly in fogs by way of flight controls and radar), or perhaps trading viewpoints between two actors, or intermittently switching between viewpoints.

Allow the Senses to Acquire Non-Veridical Content

The embodiment hypothesis states that our bodies drive our perception. In his work *Action in Perception,* Noë argues that "perceptual experience acquires content as a result of sensorimotor knowledge" (2004, 9), a claim supported by his discussion of perceptual adaptation observed in experiments which modified subjects' sensorimotor experiences. Another way to say this is that experience gets its meaning from the way our body relates to the experience. The experiments he describes used left-right reversing goggles that rendered objects on the wearer's left to appear as if they are on the right and those objects on the wearer's right to appear as if they are on the left. The goggles effectively reversed the left-right axis in the subject's relative frame of reference, creating a non-veridical situation in which the wearer's senses don't coincide with reality. Noë outlines three stages of perceptual/physical adaptation as subjects adjusted to the experience of having their left-right axis reversed. First, subjects experienced the unstable state of simple inversion of the visual field, which is inconsistent with their auditory perception (what they now see on the left is still heard on the right). In the second stage, subjects still experience the inversion, but their auditory processing yields to the information acquired by the visual system and sounds seem to also invert (sounds made on the right now seem to be on the left). The third stage of adaptation has subjects adjusting to the sensory inversion, which means that they now experience arrangements the way that they actually are when the subject is not wearing any goggles. Importantly, Noë notes, when the subjects take the goggles off, they feel the same sort of "experiential blindness" as when putting the goggles on initially. This is only one way in which visual and auditory perception are seen to emerge from our embodiment, and we have five sense organs. This concept might sound familiar, as a similar experience (among many others) was created by Carsten Höller, called *Umkehrbrille* (2001), which involves a pair of "goggles" with prismatic lenses which make the world appear upside down, rather than the left-right reversal discussed by Noë.

Motion TOOL #5

The integrated mind-body connection means that our bodies are tools that we use to think about the world. One of the primary modes of engaging the world is movement and motion in and through the world. Motion dynamics shape the way we think about the world. This is partly why the best way to learn about a new neighborhood is to walk around the neighborhood. By moving through the neighborhood on foot, you interact with that neighborhood in a physical conversation. Movement creates cumulative knowledge of the path of motion.

Motion is a basic mode of experience, it is how we encounter and engage the world. Like other aspects of embodiment, changing the way that people move will change the way they think. Perspective, attention, and viewpoint all change when motion is modulated. Motion dynamics that can be effective in shaping these three elements of experience include:

— slow movement in a large space or over long distances;
— fast movement in a large space or over long distances;
— slow movement in a small space or over short distances; and
— fast movement in a small space or over short distances

These different dynamics can be exploited to create mindfulness and intentionality, to increase observation, and to focus attention; they can also be used to hide, to distract, to bypass, and to blur. This happens because of the role of time that exists in both duration and in tempo. Slow movement takes longer because it has a drawn out tempo and because it takes more time to cover spaces that are larger. Fast movement doesn't take as long because the tempo is accelerated and the time it takes to cover spaces (even large spaces) is shortened. Slow movement can make a short distance seem long, just as fast movement can make a longer distance seem short.

Viewpoint during slow movement is heightened because there is more time to see the environment as the participant moves through it. This holds true for both participant and spectator viewpoint, although participant slow movement is more perceptible than the variations in movement experienced through spectator viewpoint. This idea that you can see more when slowing down is one of the flows of attention that you can make use of when designing for transformational experiences.

It might be useful to explore oscillations of speed in motion to blend experiences. Try these dynamics:

— oscillating speed in a large space or over long distances, and
— oscillating speed in a small space or over short distances

Switching back and forth between fast and slow movement puts control of attention in your hands as the designer. You can pair the speeds with content to create rhetorical effects in the environment that your audience experiences. You decide when to slow people down and when to move them along.

Up to this point, most of the motion discussed has been motion that the audience engages in as agents. They are moving through the environment. But other types of motion exist that do not depend on agents being in motion. Think of the difference of motion from walking along a country road vs. riding along a country road as a passenger. The views that you experience are different, but it is not as much about speed as it is about agency. When walking, the effect of motion is that of you moving through the landscape, but when riding, even though you are moving through the landscape, it seems more like the landscape is moving past you. Part of this is because you are sitting still and everything outside the window seems to be rushing past you. This is called the *parallax* effect and it is a kind of cognitive simulation (much like *fictive motion* in language) also called *frame-relative motion*. The world appears to be moving because your frame appears to be static and non-moving. The interior of the vehicle in your immediate view does not appear to change, but everything outside the vehicle is moving rapidly. It is an illusion, and as an illusion, it is a moment for hijacking the senses.

Frame-relative motion can be used to create a kind of ghost movement that persists after the visual/visceral stimulation is gone. For example, sustained and repetitive visual motion (such as a film showing trees in a forest being driven past) creates an expectation of motion that your attention scans, and when the motion stops, your attention is still anticipating motion, so the sense of motion lingers as a ghost movement.

Simulated Motion, Empathy, and Real Motion

The traces of motion in two-dimensional and three-dimensional art evoke empathetic responses that engage the body, creating sensations that mirror the muscular gestures that are implied in the work of art. Embodied simulation of these traces of muscular gestures (such as brushstrokes, knife cuts, or bodily motion) occurs as empathetic responses via a system of mirror neurons (Freedberg and Gallese 2007). If a work of art like a painting displays traces of motion from the artist, viewers can pick up on these motions and feel the sense of dynamism of a work of art as they empathetically respond to the motion of the artist through muscular simulation from the activation of mirror neurons. This doesn't exactly explain why art moves us or define what makes art aesthetically pleasing, but it does suggest that our bodies play a role in how we understand the power of art. Freedberg and Gallese argue that this happens with sculptural forms as well, since viewers experience the "felt activation of the muscles that appear to be activated in the sculpture itself" when responding to physical struggle and exertion of subjects in sculpture. In Umilta et al. (2012), the traces of goal-directed motion that shows up in abstract art are again seen to induce motor representation of the same motion in viewers' brains. These studies and others (Freedberg 2006; Battaglia et al. 2011; Sbriscia-Fiorretti et al. 2013) suggest that there are significant cognitive responses to visual image-based art, and that we experience empathetic responses to the gestures (motions) of production and content in art. The types of gestures that evoke empathetic responses have image-schematic structure that can be easily replicated in physical space. Although the studies don't address installation art, it seems likely that physical space can be organized to evoke the same empathetic effects if the composition of the installation contains the same types of image schemas that suggest motion. Image schemas can be a compositional tool when building an installation that unfolds along a path, as they can incorporate suggested motion and dynamism as elements *jut out* into the path, or *rise* or *descend,* or *follow* the path, or *pop up* here or there. You can build elements in physical space that activate the verbs of fictive motion through the composition, and by doing so, tap into empathetic responses to motion in embodied simulation.

Experience and Movement

Perceptual experience acquires content (or becomes meaningful) as a result of sensorimotor knowledge that is dependent upon our movement and the nature of objects (Noë 2004), which includes changes in the environment and the type of changes produced by our movement through that environment. Our movement through environments unfolds to create new meaning, and this opens the door for rhetorical uses of designed space through the careful sequencing of information. If movement is tied to the acquisition of knowledge about spatial environments, then the sequencing of stimuli along controlled paths should provide the artist an opportunity to structure visual and sensory information and stimuli in linear narratives. In spatial art, like installation art and architecture, this structuring takes shape as the coupling of compositional techniques with content: it is the very entanglement of form and function.

The designed physical environments of immersive art have structural composition that is characterized by dimensionality and depth. In flat images like paintings, dimensionality is suggested through compositional techniques like forced perspective, three-point perspective, and other techniques. Image composition is often judged by frameworks like the Rule of Thirds and the golden ratio (among others) to determine what is most aesthetically pleasing to the eye. But Freedberg and Gallese (2007) suggest that embodiment provides a richer and more interesting analysis that explains why we react with emotion to the composition of image-based works.

If image-based works can be explained with embodiment as viewers move their way around an image with their eyes, it seems reasonable to expect that it should also explain immersive works like installations where people, in fact, actually move through the space with their entire bodies.

Image-based works that depict spaces are representational of space, but the physical space of an installation is a direct encounter with space itself. Installation directly engages the body and the senses rather than engaging the viewer through visual approximation of sensory engagement by way of two-dimensional representation. This direct engagement and the design factor of movement through an environment enables the placement of sensory stimuli in sequences and layers that can mimic the way they are organized in real day-to-day life (like a traditional linear narrative), or they can be sequenced in ways that are fractured fragments of reality (like a non-linear narrative).

The path of movement and the manner of movement (speed, mode, etc.) through a designed space can frame the experience and construe elements of the experience in ways that heighten attention and the aesthetic value of the experience. Immersive artworks allow visitors to engage the world of the artwork through direct perception of the light, sound, scent, and other sensory conditions of the space that they move through as participants.

In the following Tools and Chapters, movement paths and manner of movement will be seen through a number of case studies to tie narrative and sensory stimuli to path-based motion through a space.

Senses TOOL #6

Visuality has dominated our encounter of the world and is typically the primary means of engaging the world and making sense of the world. Dewey (2005, 260) recognizes this primacy of vision when he states, "The organism that is set to experience in terms of touch has to be reconditioned to experience space-relations as nearly as possible in terms of the eye." Even in metaphoric structure, it is vision that is used to reason about understanding—*I see what you're saying*—although touch is close behind: *She can't grasp the concept.* Dewey argues that the visual system ought to be used to structure the haptic system, which is physical, but we already do this with metaphoric touching as Lakoff and Johnson (1999, 53–54) demonstrate with the metaphor *Seeing is Touching*: "She picked my face out of the crowd."

Historically, sight has dominated the other senses in art, but Jones (2006, 8) argues that all of the senses need to converge in the way that art produces and converses with embodied knowledge. She calls this convergence a "sensorium," defined as the way people coordinate "all of the body's perceptual and proprioceptive signals as well as the changing of the sensory envelope of the self" and describes the sensorium as "shifting, contingent, dynamic, and alive." This has a distinct dependence on the body, that "lives only in us and through us." Translate this into the hacking process by fabricating a convergence of the senses. Engineered experiences build upon the structured relationship of experiencers and their bodies by engaging the sensory faculties of those bodies and flooding the experiencer with immersive sensory data, leaving it up to the experiencer to sort out the data and make sense of their perceptions. This collection of dependencies and relationships lends itself to creating cross-sensory atmospheric experiences in which people are immersed in sensory stimuli orchestrated to affect emotional structure.

How might these cross-sensory experiences be structured? Pick any two senses and combine them to create a basic-level crossing. Make sure that the two senses can contribute to your work in terms of both form and function. Use Table 4 to help you find a pairing to experiment with in your project. After you feel like you understand what it means to blend two senses together, add in a third layer and, over time, if it serves the goals of your work rather than distracts from your goals, then you may continue to increase the number of senses you bring together. Use this cautiously, as not every designed experience benefits from crossing the sensory streams. For instance, something that is intensely scent-based might work best without crossing senses (although, it might help people concentrate if you remove other sensory stimuli through darkness and ear plugs).

Combine Stimuli from Multiple Sensory Systems

As mentioned earlier, conceptual metaphors are tools to think about one thing in terms of another thing. It is possible to create conceptual metaphors that are sensory metaphors because conceptual metaphors are grounded in bodily experience (which necessarily includes sensory experience).

You can take sensory stimuli and use them to create a compounded stimulus for an almost induced "synesthetic" experience. Although it is not genuine synesthesia, it does cross sensory stimuli to create new effects. An example of this idea might be in lighting and stage design, where light and sound are combined to get

	Visual	Auditory	Tactile	Olfactory	Taste
Visual	some feature of light (threshold, saturation, hue, color, tint, contrast) (matches/does not match) image (composition, content, subject)	image/light (matches/does not match) sound; image of a quiet scene with chaotic sound; image of a chaotic scene without sound	light (reveals/hides) texture; image suggests texture; image mimics texture; image appears textured but is smooth	image (matches/ does not match) smell	image (matches/ does not match) taste
Auditory	sound light sound image light and sound modulate in sync image and sound in sync	one sound channel cancels another sound channel; one sound channel makes an inaudible sound audible; tones conflict, tones harmonize, tones combine,	sound (matches/ does not match) sound a texture is expected to produce	specific sounds paired with specific scents	sound compliments taste; sound conflicts with mouthfeel and taste; sound descriptors applied to taste
Tactile	texture (creates/ resembles) image or visual pattern; texture interferes with light (e.g., causes shadow)	texture (produces/does not produce) expected sound	felt texture (matches/does not match) expected texture	texture shapes experience of smell (e.g.,texture holds scent, releases scent, directs scent) or specific textures paired with specific scents.	texture (matches/ does not match) expected taste
Olfactory	smell (matches/ does not match) image; smell (evokes/does not evoke) an image; or present smell with blindfold, in darkness, or void of visual stimuli	smell (evokes/ does not evoke) sound association; scent note (matches/does not match) an auditory note	smell (has/does not have) texture (e.g., a thick smell)	scent blends with scent to compound; or individual scents subtracted from a common scent composition	smell blocks taste; smell amplifies taste; smell disguises taste;
Taste	taste (matches/ does not match) color, or visual cues about texture	taste (matches/ does not match) expected sound; Also, auditory and tactile elements of taste: mouthfeel does not match sound	taste (matches/ does not match) texture (e.g., mouthfeel and flavor)	taste (matches/ does not match) scent (e.g., durian smells like “garbage” but tastes sweet)	one taste replaces another taste; one taste anticipates another taste; one taste balances another taste; on taste contrasts with another taste

Table 4. Combining the Senses.

brighter and louder in sync with each other. This synced ramping up of volume and brightness can become a tool to tap into people's emotions to excite them and draw them into the temporal rhythm of the event experience. The mood brightens and evokes new energy as the room brightens and the sound gets louder. This is a sensory metaphor where the visual and auditory stimuli (concrete elements in the experience) are used to think about emotional effects (abstract elements in the experience). The abstract effects in this case would be excitement, a brightened mood, a mood of anticipation as the lights and sound increase, and a moment of elation and release when the lights and sound reach their peak.

Sensory Experiences and Descriptive Metaphors

The types of language that people use to describe what is happening in the scene itself reveal the structure of the sensory metaphor that they experience in the scene. This is a bit tricky, but if you work backwards from the abstract effect that you want to evoke and find the image schematic structures that best suit the metaphor you used to describe the effect, you can develop a physical intervention that uses the same image schematic structures, along with sensory stimuli, to build the metaphor in physical space. This might be the most important idea in this book.

Here's an example from cognitive linguistics:

If the types of language people use to talk about events and experiences provide a set of indirect tools for designing cross-modal experiences and building multi-sensory metaphors, it should also be possible to use descriptions from an experience (like music and sonic experience) to build another experience of the same type (another musical experience) as a sort of data "visualization." It's an interpretive process of translating from one domain (the sensory domain) to a non-sensory domain so that the non-sensory information can be understood through the senses.

Antovic et al. (2013) conducted research with non-musically trained children, in which they played ten different musical statements to the children and then listened to how the children described the music. They used simple music, like a scale, or two notes played in different octaves. The children described the music and the researchers then analyzed the descriptions to see what kind of structure the descriptions had in common. Many of the descriptions fit into the pattern of a few conceptual metaphors. For instance, if a child described the sound of a changing pitch as "went high, high, high, and low, low, low" or "the first one was low and the second one was high," the researchers categorized these statements as reflecting the metaphor *Pitches are Heights*. Remember these were students with no musical training and they were also children. The description of heights had nothing to do with expert knowledge of frequency or spectral analysis of sound, so the use of height language is related to some way that the children conceptualized the sounds.

It was easy to take these ten basic musical statements and identify the metaphor and the skeletal image schemas used to build those metaphors. This means that I now have a set of tools to build a new piece of music based on those metaphors, and every time I want to suggest a particular metaphor, I know which type of sounds to use. Just because the students used similar metaphors to describe these ten pieces of music doesn't mean that other people will use the same metaphors to describe similarly constructed music. But the trend in this data suggests that at least some people will respond to the music in a particular way, and that is helpful for our purposes.

The ten pieces of music all had image-schematic structure (revisit Figure 8) that ranged from: container, up-down, center-periphery, link, part-whole, force, front-back, path, and source-path-goal.

At this point, I can take these image schemas and work with them in terms of another sensory system, like scent. By pairing the image schema from a piece of music with the introduction of a particular sequence of scents, a scentscape can be built that matches a song. Now a sensory installation can use a scent and a piece of music to suggest and reinforce the mental imagery evoked by the sensory experience.

Another approach would be to take the image schemas from the music and map them to non-sensory information. This enables us to translate non-sensory information into sensory experience. It's a way of helping people understand information with a sensory metaphor. Basically, this is a method of data-visualization, except it's not strictly visual, but open to any sensory system that can help people make sense of the data in terms of sensory stimuli. The image schemas from the music descriptions are very common, and anywhere else that you find those image schemas becomes material that you can translate into music. For example, the image-schematic structure of geologic events can overlap with the image-schematic structure of the musical descriptions, and it might be possible to map from image schemas of the musical statements to the image schemas of the geologic events to make a piece of music that articulates the motion of the landscape over geologic time.

Ever since the Big Bang, major geologic events have happened that have shaped the earth. These events have inherent motion that can be diagrammed using arrows to show the directionality of force. If you sketch out the force of those geologic events, you have a set of image schemas that you can then match with the image schemas from the musical descriptions. Geologic events have a strong element of force to them, with notions of pressure, expansion, movement, and collapse. These elements of force also have strong components of *directionality*: the sea-floor spreads *out,* mountains build *up,* striations and sedimentation pile *up,* ravines erode *down* into valleys that spread *out,* and so on.

In order to map between the image schemas of the musical statements and the image schemas of the geologic events, I listed the major events that have happened throughout geologic time and classified each event by which image schema structured which aspect of that event. The list of basic types of geologic events includes the Big Bang; formation of the moon; continental movement, break-up, and formation; oxygen catastrophe (depletion) and the oxygenation of the atmosphere (filling); orogeny (mountain building); glaciation (forming, melting, moving, scrubbing, shrinking, grooving, transporting erratics, reglaciation); volcanic activity; plate tectonics; seismic activity; icehouse earth; greenhouse earth; the rock cycle: sedimentary, metamorphic, igneous processes; paraconformity; angular unconformity; disconformity; and asteroid bombardment, meteorite impact, and subsequent crater building. Each of these geologic processes has an image-schematic structure that maps to the structure of one of the musical statements.

I then drew diagrams of the image schemas for each of the geologic events, using arrows to model motion, force, and directionality. Each of those drawings maps to a certain subset of the overall list of image schemas that link back to the musical stimuli.

What I had was a list of image schemas from the music and a list of image schemas from geology. I was able to find a third list in the overlap between the music and geology, and that third list is the mapping between the two lists, showing which pieces of music embody the motion of which geologic event, thus linking geology to music by way of correspondences between metaphorical motion and physical motion.

So now those musical statements can be used as really rough building blocks or the as the backbone of a musical skeleton that can be elaborated to produce different musical progressions that structurally mimic the motion of the geologic event. These progressions can be finessed to have sound contours that characterize different aspects of the structural geologic motion, such as speed, amplitude, and pitch, all modulated to make the music richer.

For example, limestone and dolomite form in shallow seas from fossil shell and coral through the process of sedimentation and layering. This means that a layer of stone could be represented with a layered building of sound, using the music stimuli that maps to the metaphors *Pitches are Heights* and *Structural Change*. These two metaphors occurred in six of the musical stimuli in Antovic et al. (2013), so those six stimuli provide input into how to build a more elaborate musical statement that can model the motion of the sediment being laid down layer-by-layer in a shallow sea and the subsequent increase in the height of sedimentation as the layers build upwards over time as the whole substrate solidifies into stone. In sedimentation, there is a downward force of each layer being deposited and a corresponding upward force of the whole ocean floor raising in elevation. Some kind of dominant downward sound slowly blends into a subtle upward sound, and a constant high-pitch drone is overlaid on that downward-to-upward transition, since, in terms of height, "shallow" is one of the metaphors the children used for high pitches and this process of sedimentation occurred in shallow seas.

This is just one way of mapping between the descriptions of a sensory experience and non-sensory information. It is not limited to sound, and this model can be used with any sensory experience. The point is to enable something that is not typically sensory as if it were sensory through the use of descriptive metaphors and shared image schemas. This is the structure of a multi-sensory metaphor.

You can collect language descriptions anywhere you want: it can start with your own descriptions, or descriptions people make on the news, or you can interview people about experiences and events. You could pick experiences like disasters or accidents, parties, birthing experiences, or anything that groups of people can describe that can become source material for you to study. Then take those metaphors that they use, extract the image schemas, and use the image schemas to build your sensory experience using the process outlined above.

Building Atmospheric Moods

Losing control of the body shakes off your sense of agency and the world seems to move around you while you remain stationary. This is kind of like the feeling of swimming in the ocean in playful heavy waves, having the body tossed around in the softness of water, with the force of waves as they break on the body, not being able to resist the ocean spitting you out onto dry land, not being able to resist the ocean pulling you back in again—you lose your control to the capricious agency of the ocean.

This feeling of being controlled by the ocean is much like being a part of the ocean itself. Movement in the ocean is simply waveforms of energy passing through water and the water gets caught up in the waveforms to produce what we think of as waves. This naive physics might lead down the wrong path, but it nevertheless illustrates that matter (in this case, seawater) loses its agency to the force of waveforms and this combination of waveform and water is part of what we think of when we think of the "ocean."

Loss of control to a spinning world resembles Freud's description of the "oceanic feeling"—a sensation of not being separate from the world at large, an "insoluble bond" between an individual and the world—the oceanic feeling is a feeling "of belonging inseparably to the external world as a whole" (Freud 1929). Oceanic feeling is something we experience in infancy when we can't tell the difference between our bodies and the world—everything is one and we are a part of it. Freud describes this as a time where there is no differentiation, and that the first notion of something other than the self is the encounter with a mother's breast. Whatever your views on Freud, the simplicity of this idea of the self being one with the external world is a deeply immersive feeling. Fostering experiences that conjure these atmospheric moods is one of the goals of engineering experiences —giving people oceanic experiences is part of creating wombs for personal transformation. When an audience connects to an atmosphere, they experience a moment of presence and live in the moment with openness. The atmosphere is what controls them: the atmosphere becomes the agent that controls the self.

One way to create these oceanic/atmospheric experiences is to tap into sensory systems and emotion to try to exploit sensory perception as it engages the emotions. It might be possible to specify how our sensory perceptions interplay with our emotional states. David Freedberg discusses this issue at length and admits that many people have trouble accepting that there can be a systematic (i.e., rule-based) approach to understanding how art forms evoke emotional responses. He writes that,

> Even if we assume that we may establish a syntax for the relations between how pictures look and how we cognize them, I believe that there is a further syntactical level: between the look of a picture and the emotions it arouses. And the rules for that syntax, I believe, are innate and specifiable. The general view, of course, is exactly the opposite. This more popular view holds that the emotions are not subject to reason or to any specifiable set of rules; and that very little if anything can be said about the relations between pictures and feeling that is not purely contextual or idiosyncratic. That, of course, is not a view I share. (Freedberg 2006, 83)

Freedberg takes the view that proportion and ratio give power to variation and differences in the textural qualities of art works to enable them to excite emotions in patrons. It could be that this logarithmic approach to aesthetics is another manifestation of figure-ground organization. These proportions differ based on the medium and it follows, then, that figuring out which proportions converge to evoke which emotional responses in each particular art form is necessary, in order to build a general theory of emotional cues in art. For example, Freedberg demonstrates that it is not enough to claim that there are modes in music that relate key signatures to the emotions supposedly aroused by those keys, because music is much more multifaceted than is suggested by a simple correlation of key signatures to moods. In Freedberg's estimation (2006, 86), it is intervals and proportions that provide better frameworks for the link between emotions and the structural architecture of any artistic composition. The ideas in this book are built on the idea that sequencing and timing of elements in the physical space might underpin the emotional response trigger.

Our focus is slightly different than Freedberg's approach to emotion evoking structures in 2D art, because engineering experiences often entail having multiple sensory experiences which overlap, whereas in a single-channel art form (e.g., painting, or musical composition), the emotion-evoking proportions occur in a single sensory channel. Multi-sensory experiences benefit from having multiple sensory systems contribute to the excitation of emotion, in which the sum is greater than its parts. While our model is intended to fit within Freedberg's notion of emotion-evoking proportional syntax within sensory-specific artistic composition (how a sensory system works internally to evoke emotional responses), our model benefits from a coarser approach and begins to explore the syntax at a more general level to see how sensory systems work together (externally/jointly) to evoke emotional responses.

In order to build multi-sensory experiences that have any sort of cohesion, it is important to coordinate sensory systems in the production of the emotion-evoking stimuli. As mentioned earlier, one way to create atmospheric experiences that engage people emotionally is to use sensory systems to engage emotions and evoke responses. One of the paths into emotional and sensory systems is through creating sensory metaphors by combining multiple sensory signals (e.g., visual cues and auditory signals) in a process called cross-domain mapping.

Think of cross-domain mapping like an old telephone switchboard operator, connecting one phone line to another phone line to create a telephonic link between two people. Cross-domain mapping is connecting one sensory system to another sensory system so that the two systems work together to enhance some sensory experience. Connecting sensory systems has an intensifying force that can be used to control experiences and ambient scenes. The trick is finding the connections that work for what you want to do.

Inducing "Synesthesia" (Crossing the Senses and Cross-Domain Mapping)

It is interesting to think about mapping between the different senses—creating links between one sense and the other so that an experience of one sense is enriched by a triggered experience in another sense. This compounds the experience, think again about the coupling of stage lights in a theater turning on slowly as the auditory volume of a band increases until the lights are bright and the sound is loud. The two gradations feed into each other to enhance ambient and attentional experience. This is a cross-modal mapping that is not controlled by the audience, and everyone in the audience experiences the same mapping of light and sound. When the lights and sound peak, the coordination of stimuli intensity is often experienced as a moment of flow.

Note that this cross-modal mapping is **not** an authentic form of synesthesia (the section heading is misleading), and labeling an experience "synesthetic" marginalizes individuals who experience true automatic synesthesia. Authentic synesthesia occurs in developmental processes that we presently do not bioengineer. Hubbard defines synesthesia as "an experience in which stimulation in one sensory or cognitive stream leads to associated experiences in a second, *unstimulated stream*" (Hubbard 2007, 193, emphasis mine). In synesthesia, the trigger is experienced in two sensory streams—for example, a sound (auditory system) and light (visual system). Sound does not typically trigger the visual system, so when synesthesia occurs as a result of a sound being made, the visual system would be the unstimulated stream referred to by Hubbard.

In the theater example, both auditory and visual systems *are* stimulated by changes in the environment, sound is getting louder and light is getting brighter. This is a simple pairing of sensory systems, and a coordination of the stimuli used to engage those sensory systems. You can pair any of the sensory systems through coordination of signals or even contradiction of signals (as seen in Table 4). Some pairing will result in richer experiences than others, and you will need to test which couplings work best for your context (site, materials, duration, audience expectations, etc.).

Because sensory systems can be paired and coordinated to create multifaceted perceptual experience, the overlap in sensory experience tends to be uniform across the entire audience and does not vary from one audience member to the next. This is a major difference between general sensory mapping and synesthesia, because synesthetic experience typically varies between individuals (while many individuals map sound and color, the particular associations can vary individual-to-individual). Another major difference is that the audience might not be **conscious** of the elements that trigger their experience of the cross-modal mapping. In our example of the coordinated stage lights and volume, the audience might not recognize that light and sound are coordinated. But in true cases of synesthesia, there is a consciousness of the trigger-effect pairing.

Not to beat a dead horse, but according to Auvray and Farina (2016), cross-modal mappings don't fit the definition of congenital synesthesia because they aren't **idiosyncratic** and may lack systematic **pairing** between stimulus and association as experienced over time. But cross-modal mappings can exhibit **consistency** (lights and sound could achieve consistent ambient and attentional patterns over time as participants learn the mapping), and the stimulus-triggered cross-modal mappings can be experienced **automatically**, which are features that cross-modal mappings share with synesthesia.

What it boils down to is that cross-modal mappings, while sharing similarities with synesthesia are in fact different. These differences give new understanding into what cross-modal mappings mean to artists engineering experiences which are designed to coax audience members into richer experiences by pairing sensory systems and coordinated stimuli.

For example, the comparison between cross-modal mapping and synesthesia shows that:

- cross-modal experiences are not idiosyncratic (which means they can apply generally to broad audiences and achieve desired effects);
- cross-modal experiences can be consistent over time (which means planning for repeatability is possible);
- consistent mappings can encourage learning (cf. learning, embodiment, and sensorimotor simulation metaphors like the experiments conducted by Slepian and Ambady 2014);
- cross-modal mappings are not necessarily accessible in consciousness (which means triggers can be subtle and effects can seem more magical); and
- cross-modal experiences are automatically experienced (importantly, because both sensory streams are being stimulated) which means effects can be timed in event structure permitting control of onset, peak, and fade-aways (tail).

Blocking Senses—Augmenting Senses—Altering Senses

Blocking the senses can be a powerful way to create memorable experiences because we don't often have our senses blocked in everyday life. The act of blocking stands out in attention as a figure of salience in our memories.

Consider the experience of eating a steak. If you are not a vegetarian, you will likely eat many steaks in your life. Some of them will be better than others, and in thinking back over your life of eating steaks, you may have a few particular steak moments that you use to "reconstruct" the experience of eating a steak in your mind. Most of the steaks that you eat will be forgotten, so the experience value of eating a steak is not very salient in your memory (I am using memory in a non-technical sense). If you are a chef, you want your steak to stand out and to be one that the guest remembers above all others. How might you achieve this goal? Perhaps by taking away your guest's sense of taste. While this sounds a little crazy, if you try it, the experience will stand out in the guest's memory because it is so different. How might a chef go about removing the sense experience of tasting a steak? Encapsulate it in gelatin capsules and offer a pile of capsules to the guest with a glass of water. Guests will transfer the steak and its nutrients to their stomach without ever actually tasting the steak, and every time they take a pill in the future they will remember being at your restaurant. (Note: if you try this, you have to pulverize the steak before packing it into the capsules so that the food is easier to digest, and merely cutting it into slivers will cause severe indigestion.)

Maybe you've already done the capsule steak dinner thing and you're running out of tricks. What's next? Well, blindfold the guest and give them a capsule meal that they can't taste or see. They won't even know what they ate! Another visual trick is to color all of the foods the same with food coloring. What is it like to eat a meal that is colored completely black? How will it affect the food? What does it do to appetite? Eating black food in an otherwise normal setting is different than eating food (regardless of color) in the dark. Eating in the dark prevents you from seeing anything around you. Eating colored food in an otherwise normal setting keeps your focus on the fact that it is the color of the food that is different. But this isn't new. Restaurants like Opaque in San Francisco have experimented with eating in the dark as the primary mode of dining in their restaurant. While this might seem new, experimenting with altering the experience of diners has been around since before the Italian Futurists Meals devised by the fascist Marinetti in the 1930s, and probably much earlier.

Touch is a difficult sense to eliminate because of proprioception, but you can dampen it and modify it. Consider how it feels to eat with your fingers and hands: you have direct contact with the food and feel it squish and tear under your goal-directed muscular manipulations. When you use a fork and a knife (or chopsticks), you are mediating haptic signals through the body of the utensils and you can gauge how much pressure to use to pick up, hold, or cut food on your plate and how to steady the food as you transfer it

Image 11. Marije Vogelzang, *Sharing Lunch*. Photo by Kenji Masunaga. Courtesy of the artist.

to your mouth. So if mediated touch can do nearly as good a job as direct touch with how you control food, maybe further mediating touch might effectively dampen the sense enough to create a memorable experience. Obviously the best course of action is to give your dinner guests a pair of boxing gloves to wear as they eat. Create clumsiness. Another option is to use elastic bands to create tension differentials between the two hands to limit the effective range that a person has with flexibility in the process of eating. Trick the muscle memory by drastically altering normal capacity with abnormal restrictions.

Eating designer Marije Vogelzang created a moment of haptic give and take when she arranged Shared Meal (Images 11 and 12), a meal eaten at a table with a unique tablecloth that connected each of the diners together such that by moving to reach something on the table, the action would shift the bodies of the other diners at the table. The tablecloth binds people together and the actions and table manners of your neighbor have an impact on your meal by constraining your motion. Each person at the table is constrained by the motion of the people around them, and everyone still gets to eat through careful negotiation of movement at the table.

Sound is a vital part of eating, so try blocking sound while eating. Earplugs help a little, but mastication resonates in the head. What if people heard the sound of someone else's chewing in their head, or perhaps an animal chewing?

Like all basic attention patterns, when you first encounter some sensory stimuli, the intensity of the stimuli makes it stand out, but as you continue to encounter that same stimuli, it becomes muted in the background. It moves from figure to ground and your body starts looking for the next figure to jump out. This phenomenon has been studied with eating and researchers call it *sensory-specific satiety.*

Sensory-specific satiety occurs when the person loses the pleasure of eating a particular food because of having been satiated with that particular food. It is *sensory-specific* because it has been demonstrated that variations in the sensory qualities of a food item reset a person's appetite. For example, it partially explains why after a filling meal you might have room for dessert. You have eaten your fill of dinner, but not your fill of dessert.

Minor variations in sensory qualities including the sight of food and the taste of food (Rolls et al. 1983), smells (Rolls and Rolls 1997), liquid volume (Bell et al. 2003), even beliefs and past experiences (Rolls 2009), such as beliefs about the amount of food necessary to reach satiety, how much food it previously took to reach

Image 12. Marije Vogelzang, *Sharing Lunch.* Photo by Kenji Masunaga. Courtesy of the artist.

satiety (etc.), have all shown similar satiety effects, demonstrating a type of satiety that is clearly multimodal.

The multimodality of the sensory-specific satiety effect makes satiety an interesting platform for crafting food-related experiences. Think of this as an extension of attention, as it enables taste-based attention to be modulated by the sequencing of tastes, textures, colors, etc. during the designed experience. The flexibility provided by the sensory-dependent aspects of satiety means that you can engineer an experience (e.g., a meal, a dinner party, tasting courses, etc.) to push the boundaries of tolerances for sensory-specific features by organizing meals in ways that diminish appetite for one flavor profile and increase appetite for the next flavor profile.

Chef Thomas Keller describes this kind of meal organization to increase appetite as following the law of diminishing returns. He creates an experience for a diner where they experience an interesting flavor but only enough of that flavor to surprise the diner, and not enough of the flavor to blend into the background. The portion is too small to get bored with the flavor. Keller says: "I want you to say, 'God, I wish I had just one more bite of that.' And then then next plate comes and the same thing happens, but it's a different experience, a whole new flavor and feel" (Keller 1999, 14).

The sensory-specific satiety effect blocks the senses by allowing a new flavor (the next salient figure) to enter the scene. Eating obviously provides a sensory experience, but less obviously, it provides a platform for telling a story, and done in a certain way, a meal can become a highly personalized experience. Compartmentalization and flavor separation can be presented in an experience that invites diners to choose their own path through the meal, creating a highly individualized dining experience. Using the law of diminishing returns in this way enables you to systematically use diminishing returns to block interest and redirect attention while a salient flavor profile fades away into the background and a new flavor profile begins to stand out. And this correlates with what we know about appetite—that *variety keeps the mouth interested in taking another bite.*

Image 13, 14. Detail, Maki Ueda, *Olfactory Labyrinth* (2013). Courtesy of the artist.

Designing Paths with Smells

Artist Maki Ueda creates spaces that require movement and active engagement with the space in order to experience the sensory elements of that space. Her work departs from traditional olfactory art which typically involves the passive reception of smells at fixed locations to be more of an embodied approach to "omni-directional" olfaction in which participants engage in decision making and path selection as they move through the space of the olfactory installation.

In her *Olfactory Labyrinth* (2013) (Images 13, 14, and 15), a grid of three different scents is suspended from the ceiling using oil

lamp wicks which diffuse the scent in a 20cm radius at each node in the grid.

Participants move through the grid by finding a particular scent and following it through the labyrinth almost like a dog would sniff to follow a scent trail. In fact, a workshop that accompanied the installation helped participants learn to "search the space like a dog," *Olfactory Labyrinth* gave people the opportunity to navigate a space with their noses and to not make decisions based on the other senses (vision and hearing) that we normally use in way-finding. This kind of installation is reconfigurable and different paths can be arranged by distributing the scents in new patterns in the grid.

Using Categorization to Design a Path Through a Sensory Experience: A Case Study

Categorization and compartmentalization are very basic to humans on a conceptual level. We have to categorize in order to think. Categorization is as basic as eating, if not more so, and this isn't a stretch: even *knowing that you want to eat* reveals that you can categorize between being satisfied and being hungry. The evidence that categorization isn't just conceptual but also trickles down into the design strategies we use in cooking and eating suggests strongly that our cognitive structure also structures our designed experiences of the world. **Categorization makes the act of tasting food possible.**

Since categorization is such a basic element of existence, we can find new ways to express categorization in people's taste experience. Simply finding ways to compartmentalize little bits of the meal into a kind of choose-your-own-adventure game path lets the diner decide *what* to eat *when* and *how*. It is ideas like this that stand behind the kitchen philosophy of people like Chef Grant Achatz, and his dish *Lamb 86* (2012) at Alinea proves it. It's a dish with 86 different components laid out in the style of a grid in 60 squares (like a compartmentalized box) on a pane of glass (Image 16).

What is striking about this dish is that it is possible that no two guests ever encountered the flavor profiles of the dish in quite the same order because the components are eaten in different orders, reflecting the choice path a diner makes during the table experience. *Lamb 86* is a guest-centric approach to the dining experience. The back of house brigade presents the bits and pieces of a dish to the diner, and the diner decides what story the dish will tell.

Tying all of this together, we know that categorization is basic to human experience, we know that variety in flavors and other orosensory factors increase consumption or at least preserves (and maybe increases) appetite for new flavors, and we know that compartmentalized dishes keep different food items separate, isolating flavors. When all of this comes together in a strategic approach to a dish (like *Lamb 86*), what you get is an opportunity to experience your own sorting strategy, and you determine this taxonomy at the table. It is human-centric and it is guest-centric—this is a chef engaging a basic element of human cognition by making a guest tell the story.

It might be easy to see how this holds true outside of the kitchen before we look at how cooking can tell a story. For instance, look at your cherished possessions: we can take the individual pieces and parts of our life and understand how they tell an integrated story.

When I was a little boy, my dad had a box of treasured possessions that he kept in a hidden place, but I knew where it was. Inside this box he had mementos of moments in his life that meant something to him—these possessions could tell a story about his life.

Image 15. Detail, Maki Ueda, *Olfactory Labyrinth* (2013). Courtesy of the artist.

Every once in a while he would show me the box and tell me stories about how the items in the box came to mean something to him, he told me his story by using these treasured possessions as props. I explored this concept in a project a few years ago called *Weaving Narratives,* in which I took eight compartmentalized styrene boxes and collected mementos of some hypothetical story.

When you take one of these eight boxes, you can "read" the items in the box as any number of plausible stories because you can put together the details in different ways. You bring your own story to the box, with your own background, and you read the item through the filtered lens of your own experience and history. The way I read the boxes is different than the way you read the boxes, and it's different than the way your friend would read the boxes, or your neighbor —the story is different for everyone because their past experiences determine when and how they enter the story.

When you take a box and you imagine a story of the contents of that box, it is your memories, biases, knowledge, experiences, and general exposure to the world that you are relying on to help you read and understand the story, and to make sense out of the objects presented before you. This will always vary between two different readers. For instance, in one of my boxes there is a tarnished wedding band—for people who have good experiences with marriage, this might symbolize something positive, but for people who have had bad experiences with marriage, this might be associated with negative feelings.

In a similar way, people eating the "same" dish at a restaurant will have varying reactions to it—this is like a form of *diner's relativity*—and what they experience is colored and flavored by their memories and overall life experiences. For example, for a long time I hesitated to eat oysters because when I was a boy, I picked up an oyster off of a pier at low tide and took it home and placed it on my desk on a little slab of marble. It sat there dying for two weeks (I didn't even know it was alive) and then it filled my room with the most awful smell I have ever experienced. Now I love oysters, but sadly, sometimes to this day, I'll see an oyster and it's like I can still smell that dying oyster smell and I get a little queasy.

The fact that memory shapes the way we engage food is not surprising, and everyone has some memories of favorite foods, foods they associate with moments of happiness, and foods that tell a darker story. As time goes on, we amplify those memories and they become entrenched in our identity—part of what makes you *you* is that you have the memories you have because you have the experiences you have. Food acts as a kind of distributed memory: we can build a memory over a bowl of soup and years later we can have a bowl of the same soup and it stirs up those memories. Since memories can be so tightly linked to foods, it is reasonable to argue that stories can be linked to foods as well.

So what happens when someone serves you a dish that has bits and pieces of your memory scattered throughout? How do you interpret that dish? Can you live in the moment and accept the dish for what it is, even if it stirs up bad memories? Maybe a mixture of good and bad memories builds a **memorable texture** to the experience of eating this particular dish.

Grant Achatz's *Lamb 86* is a dish that lets guests establish and drive a narrative of memorable texture, and it does this on two dimensions: first, it draws on the associative memories that guests have with specific ingredients (*Lamb 86* has 86 ingredients); second, it lets guests choose the order in which they approach those ingredients that are presented to guests on a 60-cell grid. Not only

Image 16. Matt Duckor, *Lamb 86 (alt view)* (2012). Used with permission of the photographer.

do guests encounter ingredients that may or may not trigger memories, but they get to choose the order in which they encounter those ingredients—the dish offers memory *and* control. You could say that guests get to *tell* the story and *read* the story at the same time.

Achatz brings categorization to the table for guests by giving them raw materials with which they can weave together a narrative that he helps them remember and create. You tell yourself a story when you pick and choose what to eat next. The path that you forge through the meal is a wayfinding experience where memories and fresh encounters with familiar ingredients are the landmarks that help you track your progress and keep yourself oriented. Tracing the path of a guest through the dish reveals not just the choices they make, but also the experience they had.

Maybe you go through the dish with this path:

> *Coffee, Mint, Oregano, Spring Garlic, Walnut, Red Onion, White Beans, Blueberry, Pasta, Thyme, Tamarind, Curry, Pistachio, Oats, Lemon, Rosemary, Red Pepper, Cous Cous, Madeira, Eggplant, Blackberry, Cumin, Endive, Red Wine, Cinnamon, Yogurt, Tomato, Saffron, Caraway, Smoke, Anise Hyssop, Rum, Fig, Clove, Fennel, Cherry, Sorrel, Blood Orange, Peach, Olive, Black Licorice, Apricot, Fava Bean, Almond, Artichoke, Star Anise, Carrot, Parsley, Dill, Brioche, Sambuca, Tarragon, Cilantro, Basil, Bay Leaf, Beets, Asparagus, Rhubarb, Capers, Honey.*

Or maybe it is this path:

> *Fava Bean, Madeira, Bay Leaf, Blackberry, Anise Hyssop, Dill, Honey, Cinnamon, White Beans, Cumin, Curry, Olive, Walnut, Red Onion, Artichoke, Tomato, Smoke, Fennel, Tarragon, Tamarind, Basil, Beets, Cilantro, Star Anise, Rum, Oats, Caraway, Apricot, Fig, Lemon, Red Pepper, Thyme, Coffee, Carrot, Cherry, Pasta, Pistachio, Spring Garlic, Eggplant, Clove, Peach, Cous Cous, Red Wine, Black Licorice, Parsley, Sambuca, Rhubarb, Brioche, Asparagus, Star Anise, Sorrel, Saffron, Rosemary, Blueberry, Oregano, Blood Orange, Almond, Mint, Endive, Yogurt.*

Or maybe you do something idiosyncratic and novel, where you:

- make little *ad hoc* groupings of flavors you like;
- eat in the order of which component you ate first in life ;
- group ingredients by world cuisine (Greek, Indian, Thai, et cetera) ;
- alphabetize components ;
- follow the color spectrum
- choose which to eat next based on colors you see around the room ;
- eat all of the lamb first, and then eat the little stuff (probably not) ;
- eat the things you don't like first ;
- make little recipes in your mouth with 2 or 3 ingredients at a time ;
- assign numbers to the grid and eat prime numbers first ;
- eat every other item to make a checker-board pattern ;
- write a word with the void created by the items you have eaten ;
- just eat what looks good to you ;
- don't pay attention to what you eat next ;
- eat everything with acidic pH before the basic items ;
- eat fruits, then vegetables, then nuts, then sauces, then spices ;
- follow some private taste language that makes you happy ;
- start at cell #1 and move to cell #60, in ascending order (but which corner is #1? which is #60?);
- etc.

However you chose to navigate through the dish, the fact remains that you have experienced a story that you participated in creating. Your encounters with the landmarks have been plot devices that advance the narrative of the dish through all stages of the dramatic format: *exposition, rising action, climax, falling action,* and *dénouement.*

This story is a layering of experience that is layered in *time* rather than *space*. You encounter the different layers of this narrative when you pick up the next bite (and each bite is separated by time), and you don't encounter a layering of space like you do when everything is served combined together (as opposed to being compartmentalized on the tray). And this time-based format for categorization influences your experience of the flavor profiles—you don't get this variety in flavor profiles when you bite into a cheeseburger. Your top teeth move through the bun, into the burger, and then meet your bottom teeth that have just come through the bottom bun to meet the top teeth in the middle. There is not much variation between the ways that two people eat a cheeseburger, but when two people are eating *Lamb 86,* there is so much variation in the flavor experience. Each dish results in a different story.

Yes, two guests both eat the same sum total of ingredients, but at any given point during the meal (say, just after component 43), the flavors you have deposited in your stomach add up to influence the way that you taste component 44, and unless you both have eaten in the same order, your expectation of how 44 will mingle with 1–43 will be different.

What you have here is a multifaceted narrative that is influenced by memory, perception, and identity. It is a participatory story. A narrative that is written by a kitchen brigade and maybe by your childhood encounter with a pistachio. It is a narrative that comes from the ability of the food itself to reach into your life history to find anchors for taste experiences. Encountering your meal piecemeal, each component individually helps you categorize your guest experience with a memorable texture where participation in the act of categorization becomes an act of eating, as well as an act of reading. Achatz is a prototypical story-teller: he crafts the elements of a story and then lets you join in, participate, and find your own way to own the story.

Creating Multi-Sensory Environments to Support Content

Our experience of life is typically multi-sensory. We feel comfortable in multi-sensory experiences because we routinely inhabit them. Sometimes when one or two sensory channels are heightened, we feel bombarded or overwhelmed by the senses. But most of the time we feel comfortable. Because multi-sensory experiences are familiar to typically-sensed people, the multi-sensory aspect can act as a supporting environment for some content that you want to communicate through that environment. This is often the case with dining experiences, where we talk about ambiance and environment, and it is often the case with immersive theater, where we talk about *mise-en-scène* and atmosphere. In both dining and theater, the supporting environment is there to highlight the content. In the case of dining, the atmosphere of the room highlights the content of the meal that includes an additional multi-sensory experience in the consumption of the food itself (e.g., the flavors, textures, colors, temperatures, smells, etc.). In the case of theater, the atmosphere highlights the story (i.e., the content) by creating the world for the story to inhabit and take place in, which gives the story credibility and believability, regardless of how bizarre and other worldly the storyline might be.

Think about the reasons you want to create a multi-sensory environment. What are you trying to do with the environment? How will the environment support or subtract from your content?

Creating Single-Sensory Environments to Isolate One Sense as Content

While multi-sensory experiences in the gallery are useful and enjoyable, there is something to be said for experiences that focus on a single sensory channel. Consider images (either paintings or photographs). An image can suggest sensory knowledge and it can convey bodily knowledge, but what it does best is draw you into the image through the visual channel. Letting images work this way without distraction from other senses allows an image to do what it does best. That's not to say that supplemental sensory channels in the environment ruin the experience—in some cases, activating the scene from the image in the room with lighting, background sounds, smells, and thermal conditions will bring a painting alive in a different way. But letting an image be an image, without other planned sensory interventions, helps the eye focus and absorb the image. Images are visual—they are not aural nor tactile nor typically appeal to other senses. While they may conjure sensory descriptions, they do so only through the eye along with our imagination and memory. Think about letting a work do what it does best. If you have a scent-based work, perhaps the scent is best appreciated in the dark or in an empty, windowless room. Isolate the sense you are trying to focus on to bring the audience into a moment of contemplation upon the sensory information they are receiving. Life is already multi-sensory. Multi-sensory experience is the default mode of experiencing life, and as a default mode, it also becomes the background. Isolating a particular sense brings that sense to the foreground for a moment as the salient figure set against the background of everyday life experience. It will be more memorable because it is a scarce mode of experiencing life.

TOOL #7

Emotion and Perception

> A man may have his heart filled with the blackest hatred or suspicion, or be corroded with envy or jealousy; but as these feelings do not at once lead to action, and as they commonly last for some time, they are not shown by any outward sign, excepting that a man in this state assuredly does not appear cheerful or good-tempered. If indeed these feelings break out into overt acts, rage takes place, and will be plainly exhibited. Painters can hardly portray suspicion, jealousy, envy, etc., except by the aid of accessories which tell the tale; and poets use such vague and fanciful expressions as "green-eyed jealousy." Spenser describes suspicion as "Foul, ill-favoured, and grim, under his eyebrows looking still askance," etc.; Shakespeare speaks of envy "as lean-faced in her loathsome ease"; and in another place he says, "no black envy shall make my grave"; and again as 'above pale envy's threatening reach.'
>
> —Charles Darwin

> Emotion resulting from a work of art is only of value when it is not obtained by sentimental blackmail.
>
> —Jean Cocteau

Our emotional life is overlaid onto our experience of everyday life. We experience emotions and moods as part of our everyday life and think of them as reflecting how we feel about the situations and events of our lives. Some people view emotions as mere embellishments to reality, a sort of personally indulgent behavior. Others view emotions as tools and as ways of being in the world. Still others look at emotions as mere chemical changes in the body. What's important for cognitive engineering is that emotions act as portals into the minds and bodies of the audiences that take part in designed experiences. They also act as leverage points that can be exploited to design an engineered experience with a specific ambiance and certain emotive qualities.

Charging an engineered experience with emotion-evoking triggers may seem like manipulation, but all art is manipulation. All of your art already helps people experience emotions whether or not you are aware of what those emotional responses are and whether or not you are intentional about helping people experience those emotions. Cognitive engineering should be about helping people see the world differently, about opening the world through some framework for inquiry. This approach helps you be more systematic about engaging people on emotional levels, and as a by-product, systematic approaches help create cohesion in your portfolio and give you powerful tools for being more consistent about cognitive engineering.

Evoke Responses of Some Kind

We respond to everything that we experience, and art, like everything else, evokes a range of responses that depend on the collective life experience of the viewer. The goal of hacking into people's everyday experience is to get people to respond to the stimulations that you provide to them through your work. Everything revolves around getting people to respond. People have past life experience and they come to your controlled environment with their own ideas and beliefs about the world.

Help People Build Emotional Experiences and Emotional Connections

Provide the building blocks of emotional experiences and let people build it themselves. People connect to different stimuli and respond to different triggers. Emotions develop over time as they fit into specific viewer narratives, providing for the gradual accumulation of the ingredients for an emotional experience, which makes the connection feel more organic. Cross-sensory approaches to connecting with experiencers on an emotional level should also be explored.

Most importantly, help people reflect on their emotions during an experience (or shortly after) in order to help them anchor a memory about the experience.

Basic Concepts: Emotions and Moods

How are emotions and moods connected to experience of the physical world? Sensory stimuli that we encounter in the physical world link directly to sensory perception when we are paying attention to those stimuli; in other words, we can sense things that stimulate our senses, or physical things simply stimulate our senses. For instance, we can hear (our sense) audible sounds (our experience) because our ears capture frequencies and vibrations (the physical stimuli). But are there connections between physical stimuli and emotional states?

Can a sound trigger an emotional response? Perhaps, but only generally. A specific sound does not trigger a universal specific emotion. Scientists studying culture and sound have found that particular music from one culture may have little subjective effect on the people of another culture, even when that music is ranked as highly emotional by the culture that produced it (Egermann et al. 2015). The mapping between specific sounds and emotions is cultural, not universal. But culture is learned, which means emotional responses to sounds are also learned, which means you might be able to teach your audience to emotionally respond to certain sounds, or you might find a way to hijack emotions your audience has already learned to respond to with certain emotions. Egermann et al. suggest that there may be universal reactions to particular musical elements (like tempo and pitch), allowing music to engage physiological response mechanisms. In their study, musical elements like tempo and pitch did evoke responses of general arousal across cultures, but it is unclear as to whether this response resulted from emotional contagion or a physiological reflex synchronization with the rhythm of the music.

This question about connections between stimuli and subjective emotions beyond mere arousal and the entrainment of heart rates to musical rhythms is a tricky question. Answers to this question depend on a number of factors and specific theoretical assumptions. Factors such as personality, cultural background, mood, situational context, relationships, individual expectations, and timing all contribute to the onset and shift of emotion from one state to the next. Charles Bukowski's poem "The Shoelace" illuminates the capriciousness of events as reasonable triggers for emotional responses: "it's not the large things that send a man to the madhouse...no, it's the continuing series of small tragedies that send a man to the madhouse...not the death of his love but a shoelace that snaps with no time left." We just can't predict what stimuli will cause a particular emotional response with any sort of accuracy because people are different and stimuli affect people differently. Even the same stimuli presented to the same person at different points in time may evoke different responses.

While we don't know the precise connections between the physical world and the emotional world, there have been attempts at connecting the two in rather general ways. Broad generalities that make sense on paper may or may not work in real life, but it is worth exploring this as a means for attempting to hack emotional states as long as you recognize that all you are doing is setting up the conditions to let people decide how much they are willing to let you control. You craft the experience, and the participants decide how to respond. Let's look at a simple model of the flow of stimuli and the resulting emotional and behavioral responses.

Robinson (2009) draws a connection between feelings and emotions by showing how they are both mental experiences that are triggered by some stimuli, and how those experiences in turn motivate behavioral responses. So, feelings like hunger and satisfaction follow this flow: If you are hungry (feeling), then it must be because you lack nutrients (stimuli) and you need food, and therefore you find food and eat it (behavior). When you are satisfied (feeling), then you are not hungry anymore because your body indicates that it has what it needed (stimuli) and so you stop eating (behavior).

This seems simple enough for basic feelings, but emotions are more complex, although Robinson argues that they follow a similar pattern of stimuli or conditions triggering an emotion that motivates some behavior. For example, returning to the example of hunger and satisfaction, we can look at positive and negative emotions in the decisions we make about how to satisfy our hunger. Let's say that years ago you ate a bad mushroom and it made you sick, and now you can't eat mushrooms without experiencing aversion (a negative emotion), and unfortunately, even though you are hungry, all there is to eat is mushroom pizza, and just the sight of the mushrooms makes you gag (visual stimuli), and so you pass on the pizza (behavior) and stay hungry, but you don't get sick (a positive consequence). You also reinforce aversion and it becomes a learned behavior (in most cases, a neutral consequence). But let's say that you are so hungry that you overcome the aversion to mushrooms and decide to eat the pizza. You take a bite, and something surprises you and you are relieved (positive emotion), because the mushroom doesn't taste the way you expected it to. Rather, it tastes good (stimuli) and so you take another bite cautiously (behavior), and before you know it, you have eaten the whole pizza and now you are no longer hungry, but you are satisfied. You may even eat another mushroom pizza at a later date (consequence).

Much of the stimuli for Robinson's lists of emotions depend on subjective experience (expectations, personal beliefs, values, etc.). Some of the emotions seem to depend on objective perceptual experience as triggering stimuli (physical responses like pain, attention concepts like sensitivity to intensity of stimuli and novelty of stimuli, and cues about good or bad environmental conditions).

It might be helpful to shift the conversation back toward attention for a moment. It is generally accepted that emotions and emotional changes are different from moods in that moods are more stable and persist longer (i.e., ground) while emotions change more frequently (i.e., figure), punctuating moods over time. These emotional-figures stand out from the mood-grounds. Since we know the difficulty of evoking specific emotional responses from a diverse audience, it makes sense to focus our energy on evoking specific moods as a baseline and permit the audience to activate the emotions that come naturally to them. If we step back and work on creating ambient environments that shape the overall mood in an event, we can invite people into creating their own emotional experience while controlling the mood as best as we can. People then take ownership of the experience and we help people build emotional experiences, emotional connections, and memories.

Conceptual metaphors can help cue the abstract experiences of audience members in the hack-space, and perhaps they will also provide a way to work with ambient stimuli in bodily experience with the structure of mood and minor emotional effects.

Conceptual Metaphors and Emotions

It might be useful to look at typical descriptions of the experience of some of these specific emotions to see how the semantics of the emotional descriptors give shape to the types of variables and parameters useful in creating those emotions (Table 5). Writers, set designers, and craft workers building cinematic *mise-en-scène* utilize emotions and moods all the time in their work and they bring together multiple sensory stimuli to produce emotional effects for audiences. Learn from them. If something you experience (like a film or a play or a song or any other designed/created work) evokes an emotional response from you, then dissect the experience and try to figure out what made it work. See if you can identify which triggers activated which emotions for you, it is likely that they will be organized as a story.

Some emotions are more complex and simple metaphors do not always map cleanly onto an experiential basis. Still other emotions have multiple metaphors that can describe facets of the emotion.

emotion	neutral conceptual metaphor systems applied to a particular emotion	how it shows up in our language	triggers in the physical world
Love	Affection is Warmth	he had a warm heart, give a warm welcome, turn up the heat in our relationship, you're warming up to me, you know how to melt my heart	warm temperature, friendly & personal interaction
Joy	Happy is Up	that was uplifting, what a rousing experience	bright light, high notes of scents, up beat and high tempo music
Surprise	Surprises are Ambushes	it caught me off guard, I wasn't expecting it, it just came out of nowhere, it snuck up on me, I was blown away	sudden changes
Anger	Anger is a Hot Liquid	he couldn't contain his rage, he was fuming, she was boiling inside	a rattling pot on a stove that is boiling over can be anthropomorphized as being angry
Frustration	Difficulties are Burdens,	I'm not able to carry on much longer, it's a heavy load this semester,	heavy objects that must be moved
	Life is a Journey	missed an exit, he took a turn for the worse, we took a detour for a few years but now we're back on track	

Table 5. Connecting the World to Language with Metaphor Systems.

For instance, *Anger is a Hot Liquid* is one conceptual metaphor that leads to descriptions like: *he was boiling with rage, his anger was spilling over into his professional life, he was steaming, she almost blew her top.* Obviously these are descriptions of physical states that we can't create directly because it would be unsafe and harmful, but conceptual metaphors help frame the problem, and a careful use of text (whether through signage or scripts) can include metaphors for anger and the network of words and phrases that fit the metaphor in question.

In the conceptual metaphor *Emotional Reaction is Feeling,* the abstract world of emotions is paired with the physical experience of tactile sensation. *Emotional Reaction is Feeling* lets us say things like "*I feel angry,*" where "angry" doesn't have a texture because it is not a concrete object with a surface. But what if you could create a texture that evoked particular emotions like anger and fear?

Social-Emotional Contagion

One way to alter the environmental mood is to seed an experience with actors who help create a particular mood. In the world of stage magic, a person in the audience who is part of the act is called a confederate. Confederates are in on the secrets the magician has even though they pretend to be a random person in the audience. If the audience trusts the confederate, the confederate gives the illusion more credibility than if the audience suspects the confederate. Confederates are always part of the trick and they make sure the trick works.

In the same way, you can design an experience that makes use of confederates to help create a particular emotional environment by using emotional contagion. Your confederates need to be good actors because for emotional contagion to be successful, it must happen unconsciously. Your confederates need to be able to convey non-verbal signals that work to evoke the type of response you want to achieve in the audience. Negative signals stand out when the background is positive and positive signals stand out when the background is negative.

Signal strength is another factor, and signals need to be subtle for your audience to adopt them unconsciously, but signals also need to be strong enough to draw the audience to harmonize their emotional state with the rhetorical emotions of the confederate. Set the tone of the background and then inject confederates to play against the background. A benefit to using confederates is that they can adapt to the audience and can modulate the non-verbal signals they are giving to match the needs of the audience. Find confederates that are good at reading people and that are believable.

To make things easier on yourself, ensure that the confederates know how they are supposed to react to the audience. Help them anticipate a range of audience behaviors and tell them how to respond to each anticipated behavior. Plan confederate responses for resistant, engaging, stubborn, or reluctant audience members. Consider also instructing the audience to engage each other (including strangers) during the experience and to be open to possibilities and encounters throughout the experience. Build openness into the audience to prime them for confederate interaction.

Another way to use confederates is to use a script that is based on a conceptual metaphor that is grounded in a bodily experience. Make the bodily experience explicit in the interaction between confederates and audience members. Then support that metaphor system by structuring the visual and sensory aspects of the experience in ways that directly reinforce the metaphor. Also consider using language that makes use of the metaphor system in some way, in order to prime people for experiencing the metaphor.

Sensory-Emotional Pairings

Take a look at Table 6 to see some positive and negative responses to sensory stimuli.

Notice how the same stimuli can often be associated with positive and negative responses, and conflicting stimuli can be associated with the same type of response. This is because emotional pairings between particular sensory stimuli are open to content, cultural learning, and context. This explains why some people love the kind of music you hate, why some people don't like your favorite food, why not everyone likes the same perfume. These differences are a kind of social variation that make life interesting. It is this variation that contributes to making inquiry-based works feel more engaging to the members of your audience, because an inquiry-based work puts the audience member's interpretation in focus.

As we've seen elsewhere in this book, it is useful to use image schemas and conceptual metaphors as tools to structure your stimuli. Use repetition, frequency, rhythm, extremes, expectation, knowledge, and size to manipulate a stimulus into being associated with an emotional response. Think about walking through a haunted house. The rhythm of the frightening encounters with ghosts is paced out such that you come to anticipate a frightening encounter based on the rhythm of encounters that you have experienced from the time you walked in the front door. Each haunted house teaches you how to be afraid of what they have to scare you with by establishing the flow of fear. Your goal is to teach your audience the appropriate emotional reaction for your stimuli and to train them when to expect to have that reaction.

	Positive Responses	Negative Responses
Visual	• bright (awaken) • dark (comfort) • strobing light (activate) • complex visual scene (engage) • minimal visual scene (soothe / peace)	• too bright (blind) • too dark (blind) • strobing, pulsing light (irritate) • visual clutter, chaotic scene (overwhelm) • barren, empty scene (frighten, isolate)
Auditory	• repetitive noise (rhythm, ambient) • unpredictable (improvisation) • arhythmic noise (natural)	• repetitive noise (annoy) • soft distant continual noise (irritate) • loud unexpected noise (shock) • unpredictable & arhythmic noise (wrong)
Tactile	• smooth, soft, clean feeling, warm (comfort) • tickle (happiness)	• rough, gritty texture, dirty feeling, cold, wet, sticky (knowledge of what you are touching, lack of knowledge about what you are touching)
Olfactory	• "good" smells (overwhelm - delight) • sweet smells (entice)	• acrid smells (overwhelm - disgust) • putrid smells (repel) • pervasive & overpowering smells (deaden sensitivity)
Taste	• tastes sweet (enjoy) • tastes bitter (savor)	• tastes excessively sweet (sicken) • tastes excessively bitter (repulse)
Temporal	• time moves slowly (luxuriate) • time moves fast (exhilarate)	• time moves too slowly (bore) • time moves too fast (frustrate)
Spatial	• large open spaces (spacious) • small enclosed spaces (comfort)	• space is too big (swallow, disorient) • space is tight or too small (constrict)

Table 6. Sensory-Emotional Pairings.

Haptic-Emotional Connections: Using Materials to Evoke Responses

Textures like other sensory stimuli can act as portals into emotion and sometimes memory. Textures have a temporality and distribution that is capricious—different objects can have similar textures and those objects can be distributed in different parts of the world. Sometimes an object develops a texture over time and so, at an early stage of its life, it may be rough, and at a later stage, smooth, or the opposite. The feel of a particular couch from your grandma's house in your American childhood might not show up again until you try on a blazer at a boutique in England. We have visceral responses to texture, especially to textures in foods—many neurotypical people can't eat mushrooms or fish or eggplant because of the texture.

I recently purchased a handbound artist book of haptic poetry created by experimental poet Brandon Stroud (Image 17). It is a disturbing book of texture poems where each page is a "poem" with a title that is meant to interpret the texture on the page. Stroud described this book as a book about rape and murder, but the textures are very ordinary taken in isolation, and only by juxtaposition and the careful sequencing of labels does the darkness of the story emerge.

The story is told through ten pages of textures taken together in a sequence that moves from inviting and comforting textures to harsh and brutal textures. All of the textures come from everyday materials in domestic life.

The cover of the book features a large blue dish-washing sponge with the label: "Home," the next story page is covered in white terry cloth, and turning the page reveals the text "gag." The subsequent pages feature sticky paper (capture), distressed wood and metal elements (trapped), a nylon mesh screen (numb), a panel of dry moss (escape), a plastic sheet with raised diagonal lines that move from left to right and top to bottom (lost), a page of hair (beautiful), a page of braille (rape), a page of chicken wire and roofing shingles (mutilation), a page of translucent white rubber shower curtain (silence), and finally, a shiny black metal surface (autopsy).

What is striking about this work is that many of the textures are everyday textures that do not necessarily have negative connotations on their own, but when they are all strung together and read-

ers are provided with the single-word descriptions that construe the texture with violence, the textures tell a haunting story and benign everyday textures become brutal and malicious and abusive.

Encountering these textures individually elsewhere in life might not evoke the same visceral responses, but when they are recruited for a tale of violence, touching the textures is revolting, and merely looking at the textures evokes disgust and chills. In fact, for my first "reading" of the book, I knew nothing of the story and intentionally did not read the labels, and it was a pleasant experience to read the texture, but on my second reading, when I knew the narrative that strung the textures together, that feeling of pleasantness disappeared and I was deeply repulsed by the story.

The story that ties the pages together drives the interpretation of those textures as disgusting. In isolation, and without labels, none of the textures are unpleasant. In isolation, some of the textures feel peaceful and comforting, but when they come together in a story of pain, they drip with horror.

Materials and textures have tight relationships with emotions when they are woven together in a narrative experience. In the same way that they can communicate violence, fear, terror, and uncertainty, they can also communicate safety, comfort, and peace. Texture is neutral until the polarity of some narrative or experience gives it a more specific meaning. Terry cloth always felt comfortable to me, reminding me of bathrobes and warm dry towels, until I read Stroud's description "gag" and immediately felt transported to the feeling of being the victim and receptor of the cruel use of a typically comforting material. This is not a trivial statement: *context drives meaning.* Map your use of texture to a narrative to harness story as an interpretive lens for texture.

Image 17. *Home,* a haptic poetry book by Brandon Stroud. Courtesy of the artist.

5

Story, World Building, and Belief

Blending Concepts to Tell a Story

A young mermaid has a problem. She's in love with a human prince but she can't be with him because she isn't a human—she is a mermaid stuck in the water. To solve this problem, she seeks assistance from a witch who temporarily gives her legs in exchange for her voice and (depending on which adaptation of Hans Christian Andersen's story you read) if the mermaid can get the prince to kiss her, then she can keep her legs for good. Now her problem is that she can walk on land, but she can't talk, and the story keeps unfolding.

All of this story unfolds naturally because of one detail: mermaids are a blend of a woman and a fish. The entire plot of this story extends from this single fact, from the exposition to the rising action to the climax and the resolution. Each plot-advancing mechanism ties directly to the details of this hybrid blend.

Conceptual blending (Fauconnier and Turner 1998) brings insight into this story and gives us a framework we can use to tell our own stories through our engineered experiences. In traditional conceptual blends, there are two inputs that blend together to produce an output. The blended concept has some of the traits of each of the inputs, but not all of the traits carry into the blended space (just like you don't inherit all of the traits of your mother and father). To get a little more technical, blends have two different types of input spaces: a regular input space and a reference space. The reference space is the dominant space that drives the structure of the blend.

We can think of the Little Mermaid as a blend with the input space of woman and a reference space of fish blended into a mermaid (Figure 12). This sounds simple. Vital relations between the input and reference spaces carry traits down into the blend and our mermaid has the characteristics of those relations. Surprisingly, in this case, the elements from the input and reference spaces that do not make it into the blend now become plot building points. The things that don't blend work to advance the narrative. For instance, in the mermaid blend, the little mermaid does not have legs (because mermaids don't have legs), and legs did not carry down into the blend from the woman input space. This fact that she does not have legs becomes a problem for her, and her limitations are what challenge her in her own life story. Legs do not carry into the blend (unless the blend is Magritte's *Collective Invention,* in which case you end up with a different blend and an altogether different storyline) and their absence advances the plot.

This leglessness becomes a point for interesting things to happen in the story, and Hans Christian Andersen was probably thinking: "Okay, so she doesn't have legs, so I'm totally going to put her in a situation where she can only succeed if she does have legs." The point is that certain elements that *don't* blend into the character of the story can be exploited later to make the story fit with the situational limitations of the character. And certain elements that *do* blend into the character provide character development in the narrative as well as the footholds for plot advancement that character traits often afford.

In designing experiences that have narrative elements to them, the way to make the story relevant to the audience is to enable members of the audience to see themselves in the story by letting them blend themselves into the story (think about role-playing games and first-person shooter games). Then, as the story unfolds, each audience member experiences the story in different ways because their inputs to the blend are unique to themselves and the resulting blend is an idiosyncratic match to that individual because of their life experience and how they interpret the story based on that life experience. If you've ever shopped at an IKEA, you understand how a retail space can blend you into some experience.

When you walk into an IKEA and climb the stairs to the display area, you walk into a domestic environment designed very much like a house in which all of the rooms are decorated like film sets, catalog displays, or even like your own house. You walk through the entire showroom, passing through living rooms and kitchens and bedrooms and bathrooms, the whole time imagining what they would look like in your house.

IKEA blends the real experience of shopping with the perceived experience of ownership and habitation. When you stand in one of the kitchens at IKEA, you are experiencing a possible future *as if* it is already present—the simulated kitchen performs a sort of compressed time travel that lets you experience the future for a few moments to help you see how you will be happy when this kitchen you are standing in is in your house. It blends the retail space with the domestic space, momentarily ascribing ownership of the kitchen to you and letting you elaborate on that idea in your own mind. Perhaps you think about entertaining guests and you begin to let your mind tell you a story about your life if only you had that kitchen. And it isn't just you: other shoppers also blend themselves in with these same display scenes, so that everyone is collectively imagining how these showrooms can be part of their lives, and it is an effective sales tactic for IKEA.

Building experiences that feature blends gives audiences a place to enter the story. Each individual in the audience will experience their own version of the blend, and they do the work to contextualize your story within their own life experience. You'll have to figure out where you will build doorways into the narrative, and you should also provide multiple entry points, but let the audience actually open the doors themselves. The audience can be asked a direct question about how your engineered experience fits with their lives or they can be subtly led into viewing how it fits with their lives by creating a scene that is relatable. Throughout the experience, the audience can be given choices and their preferences will be amplified in the story and experience, especially if emotive content and empathy-inducing elements build portions of the experience.

If something is believable and it is something that people can relate to, they will experience moments of empathy in which they can share other individual's feelings. When I read Hemingway's "Hunger Was Good Discipline," I entered the literature empathetically because I know what it is like to walk around a city and I know what it is like to be hungry, and even more, I know what it is like to walk around a city while I am hungry, and what it means to smell food and be unable to have the food that I am smelling (smells wafting from restaurants, summer grilling in a park). So I join Hemingway as he walks around Paris trying to avoid good-smelling food throughout the day. My experience of the world blends with Hemingway's described experience in reading his story, and because I can relate to it, his story has more meaning for me.

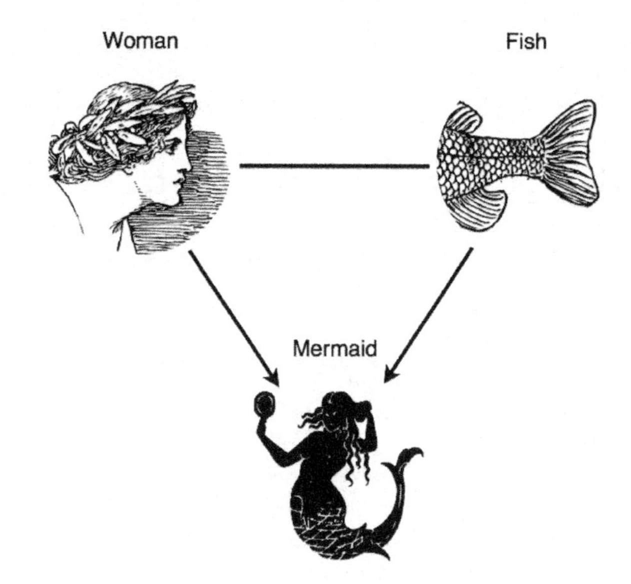

Figure 12. Mermaid Blend.

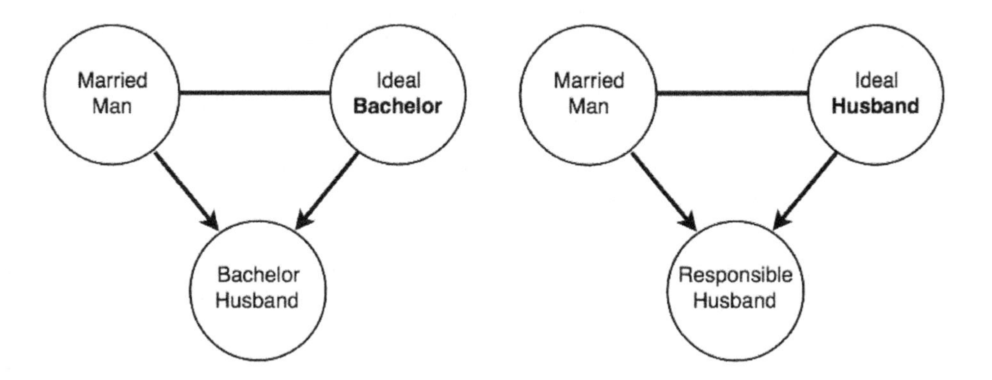

Figure 13. Bachelor Husband Blend.

Everyday Blends

Blends are not necessarily extreme hybrids of two things like fish and women; they can be banal, like a blend of a married man in one input space and a bachelor in another input space (Figure 13). In the narrative that extends from this blend, there is a man who is married but still acts like a bachelor or who has loutish, prototypical bachelor desires despite being in a committed relationship. If these behaviors manifest themselves in his marriage, they potentially create strife between his spouse and his desires, or they build frustration in his spouse, or he neglects his spouse, or perhaps he is unfaithful to their exclusivity, etc. All of these points of tension can turn the arc of the plot in new directions, and all of those tensions stem from what was or wasn't carried into the blend.

In critiques of blending (Gibbs 2000), one of the primary concerns is that blends happen everywhere and so they don't have the ability to explain something unique about cognition—if everything is a blend, then blends are meaningless to science because they don't show us something we don't already know in a way that is falsifiable and which can't be explained by other theories. Conceptual blending in this view can only be descriptive, and not explanatory. Gibbs' description-only view of blends led me to begin to think about how I could use blends not just descriptively, but *productively,* as tools of creation that can model change, progression, and feedback loops. In this new productive view, blends seem more like a *process engine* ready to be harnessed into a process of combinatorial creativity.

Modeling Change and Narrative Progression with Blends

A narrative can be centered around a single blend, or it may be a combination of different blends, or it may be a sequential progression of blends that refine the original blend like a series of feedback loops. Blends also fit into temporal timelines and can chain together to model various changes in the narrative. In the bachelor-husband model, ideals from bachelor life were not discarded nor unlearned as he moved through his various roles as boyfriend, fiancé, and spouse, so he retains his bachelor traits. However, if something in the bachelor-husband's life causes him to rethink his behavior, the blend can change.

Perhaps something happens in the bachelor-husband's experience to cause him to reevaluate his behavior (it could be the consequence of one of the points of tension from something not carried into the blend), and he changes his behavior during this learning experience—perhaps he seeks advice from a counselor who helps him see that his behavior as bachelor-husband needs a different benchmark (e.g., a role-model-husband). At this point, bachelor-husband (with all of his baggage) is the input space, and role-model-husband is the reference space, and a new blend emerges that combines the two, and maybe in this blend, the bachelor traits don't carry into the new blend and the fact that the bachelor traits don't carry into the new blend represents a transformation in the story. The blend models the progression of a learned experience. If the original blend is changed so that the ideal husband is now the reference space and the married man is now the input space, then the story that emerges is different altogether—that of a "responsible" husband. The point of this illustration is that blends change, they adapt to new information, they can learn and progress and evolve, they always generate based on the inputs to the blending process, and the resulting blend often is greater than the sum of its inputs. This adaptation capacity makes blends useful for dynamic experiences and user-driven narratives.

Questions for building blends into your experience:

— What is the overall story of the experience you are engineering?
— How do other people enter the experience?
— Will you have people bring their own emotions or knowledge to the experience?
— What relationships do you intend to form between the input brought from audience members and the reference space provided by the structure engineered into your experience?
— Which relations carry into the blend? For example, as people try to relate to your experience, which things will you permit them to carry into the experience? For example, your experience may require people to focus on the prototypical model of motherhood (birth mother) without thinking about other models of motherhood (nurturance, adoptive, biological mother, donor mother, genetic mother, etc.). Your audience will have varying experiences with their own mothers, some of them will not know their birth mother, some of them might know their genetic mother, but not their birth mother (as is the case with many surrogate pregnancies), some of them might consider a woman other than their birth mother to be the woman who nurtured them as if she were the mother. Since motherhood as a concept has many different models, it is important to clarify the model you are using in your blend. This applies to any concept with multiple models.
— How do the non-inherited elements contribute to plot development in the experience/story?
— Does the audience blend with the experience?
— If you are building an experience that blends two different ideas, how does the blend tell your story for you?

Designing Blends for Your Experience

It might seem difficult at first to design an experience around a blend, but think about it this way: **you can turn a blend into a story, and it's easier to think about designing an experience around a story.** Because a blend can turn into a story that ties the experience together into a cohesive experiential narrative, it is useful to at least play around with how your concepts might be turned into blends, or how they might already be blends of two or more concepts. The way blends have been present here is slightly different than their traditional role in cognitive science which uses blends to *analyze* the world. In this context blends are being used productively to *create* the world. In this way, it is a kind of combinatorial creativity and it becomes a simple engine for a kind of generative art where the story line is the generated product.

In their most basic form, blends are combinations of elements that produce something novel—sometimes the blend is mundane (like a fork/spoon spork, or a retail space like IKEA) and other times the blend is pyrotechnic (like a mermaid, or an atheist megachurch). How you bring blends into the structure of your experience is up to you, but their inclusion should always be motivated by how they serve the story.

It is always good, and perhaps unavoidable, to have your audience members be one of the input spaces in the blend. They bring their identities and context to the designed experience and the act of engaging in the experience itself can become a simple blend. They come as inputs, you provide another input, and you help them create mappings between their experience and the input you want to blend with them, and as the experience progresses, they carry some of their own details and relationships into the blend with some of the details and relationships you have provided. As people continue on in their day-to-day lives, the blend created by your engineered experience will manifest itself in how those people encounter elements of their existence apart from the engineered experience. They have been changed through the process of blending with your intentions in the experience. For instance, maybe someone thinks about owning a drawer organizer every time they open their kitchen drawer because they saw one in the IKEA showroom. The next time they go to IKEA, they buy one and organize their drawer. IKEA blended organization into the story of what it means to be at home in a kitchen and when that person went home to an unorganized kitchen drawer, they were struck by their lack of organization in a fresh way. This leads to a discrete action, and IKEA made a long sale because they sold you the idea of organization during your previous visitor experience in their blended showroom.

To get you started on thinking about ways to blend experiences for people, consider the following list of some basic possible combinations that you can craft. In order to create a blend, find some combination of the familiar and unfamiliar and start thinking about the possibilities. You can:

- provide familiar experiences (patron past experience blends with familiar experience);
- provide unfamiliar experiences (new to patron);
- provide unfamiliar experiences in familiar settings;
- provide familiar experience in unfamiliar settings;
- bring together familiar elements;
- bring together unfamiliar elements;
- bring together a familiar element and an unfamiliar element; and/or
- bring together elements that are different along some dimension.

If you provide a familiar experience, patrons know the general script of how they are supposed to behave. This provides a kind of comfort that might help to disarm patrons and drive them towards moments of openness. The IKEA showroom experience is familiar because it is laid out as a domestic space. Watch people in the store and you will see that they have no inhibition of "playing house" in the showroom. People will lie in beds, sit around dining room tables, pretend to cook, pretend to shower, and sometimes they will actually groom themselves in bathroom mirrors.

No one does this sort of thing in traditional furniture stores. The difference is the tight cohesion of the familiar domestic space offered in IKEA showrooms: this familiarity is comfortable and people let their guards down and move toward moments of openness where they try out products in highly personalized ways to see if they like them. IKEA creates openness in people by making them feel at home, and that openness results in more sales.

If you want to provide a ceremonial or ritualistic experience, choose between providing a familiar experience in an unfamiliar setting and an unfamiliar experience in a familiar setting. Consistency and formulaic order are components of ritual structure that separate the organization and intentionality of the ritual act from the way that we approach the rest of our everyday lives. Ritual is a practice that brings new meaning and order to the world of the everyday experience, and blends of the familiar and the unfamiliar can bring a ritualistic approach to an experience. A familiar experience in an unfamiliar setting rips the familiar practice from its context and places it in a new setting, which can make the familiar experience seem out of place and otherworldly. On the other hand, an unfamiliar practice in a familiar setting heightens attention to the new practice and may evoke states of mindfulness to the newly unfamiliar practice. For example, following a recipe for the first time often elicits mindful behavior with thorough attention to the precision of measurements and the temporal sequencing of activities. Although not an exact match, in many respects this attentiveness to details while cooking a new recipe looks very much like a ceremonial practice.

Other types of blend experiences might include giving people the experience of living somewhere else, of having a different sort of body, of having the sensory abilities of a particular animal, of having the viewpoint of another patron, of having secret knowledge

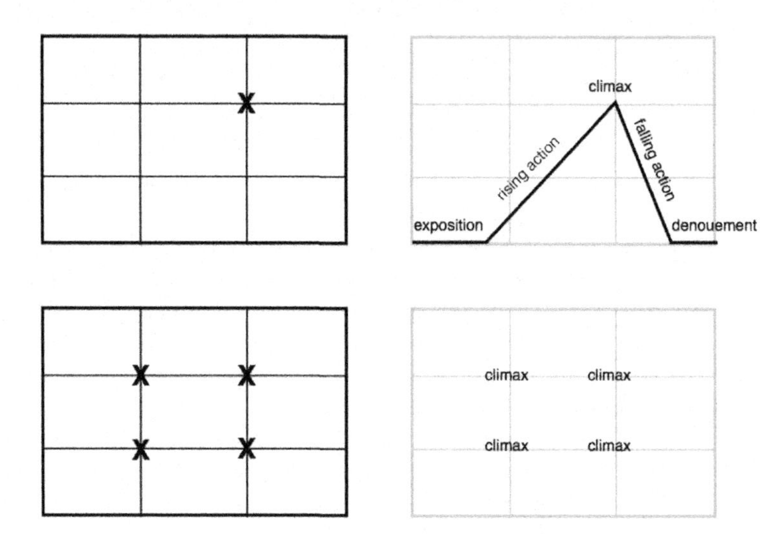

about someone else (voyeurism, surveillance, monitoring), assuming a different lifestyle or standard of living, transportation through time (past & future), and switching roles. This is not an exhaustive list of ideas, because in the combinatorial blend, anything can be blended with anything else. The better blends for engineered experiences may be the ones that feature tighter integration between patron and the engineered experience, because they weave the patron into a tighter relationship to the plot structure of the experience story.

Entry Points in Story

The audience needs to be able to enter the story at some point. Building portals into the story will let the audience blend with your designed story, creating their own experiential blend. They will bring the context of their own lives into this experience blend. The touch-points for entry into the narrative could be as simple as giving people a choice as to where they start off in the story, or which character's viewpoint they want to inhabit, or letting them choose a set of words that they relate to as a way of helping identify which character they might best empathize with, and then letting them enter the story world.

Basic Plots and Baking Story into the Experience

People recount their experiences to each other by telling stories. Stories are the medium for memories and they are the tool for sharing and communicating. If you want your experience to stand out and become a memory, try pre-scripting the experience to be easily told as a story. You can help people understand your experience if you help them experience it as a story. Experiences that make good memories also make good stories, and it is easier for people to tell stories about something that they experience in story form because you've pre-loaded the memory and the telling with the narrative arc or plot that you have chosen as the framework of your designed experience.

"The intelligent read, the wise read literature." Supposedly Edgar Allen Poe wrote that, and the lesson I take away from this quote is that reading literature might give you a better chance at understanding how to make something meaningful, because you have better insight into the experiences of life as told in traditional narrative form.

With this in mind, learning about plots and learning about story are important parts of designing experiences. If you do not understand story, search the internet for examples of plots in film and literature and then watch those films and read the books to

Figure 14. Rule of Thirds and the Narrative Arc.

see how plots unfold. You do not have to be a literary critic to design an experience, but it does help to know how plot can be used as a tool for storytelling. You may already have an intuitive sense of plots and may recognize plots by their names. Booker (2004) outlines what he believes are the seven basic plots: *overcoming the monster, rags to riches, the quest, voyage and return, comedy, tragedy, rebirth.* Tobias (1993) lays out twenty basic plots, including: *quest, adventure, pursuit, rescue, escape, revenge, the riddle, rivalry, underdog, temptation, metamorphosis, transformation, maturation, love, forbidden love, sacrifice, discovery, wretched excess, ascension,* and *descension* (and at the end of each chapter he provides checklists to use when writing these plots, which are useful in building stories that traditionally work well). Other researchers claim there are upwards of sixty basic plots. While you might know the general trajectory of a plot structure by looking at its name, do a little research and build lists of events that can happen in the specific plot type so you can see what kinds of twists and turns you can use in the storyline of your own designed experience.

If plots don't seem to work for the story you are trying build into your experience, try to at least structure your story around the basic elements of drama (Burke 1969): five simple parts of story rhetoric that make it easy to break up the world of your engineered experience. The elements are *act* (what), *scene* (where/when), *agent* (who), *agency* (how), and *purpose* (why). They are similar to Spradley's descriptive question matrix (Table 1) and account for the same content. Earlier, the domains were presented to help you figure out how to break up the experience into manageable layers. These dramatic elements and plot structures are presented to help you figure out how to make a story happen in the space, with the participants, over a stretch of time, toward some goal, using some story that ties it all together. This is a focused view on building story into the design of the experience up front.

Think about the general structure of traditional narrative: it is kind of like a linguistic Rule of Thirds. In the same way that an object of importance is highlighted in one of the points of salience in a photographic image, the climax of many traditional narratives happens two-thirds of the way through the book. Climax fits a third of the way through (Figure 14). On the surface, discourse structure and spatial structure share a pattern of salience. Nontraditional linear narratives enable you to move information around, placing the salient portions in other thirds — early or late, top or bottom — and then filling in the details according to the arc. Note that this is not the same as a non-linear structure, which plays with chronological ordering by creating disjunctions in time.

Storytelling through Oscillation: A Case Study

I remember the eruption of Mount St. Helens in 1980. As I watched the aftermath on a television screen, I did not yet know what I was seeing. Everything looked gray. Everything looked dirty. I did not know what pyroclastic ash was, and I did not know what a volcano was. I was not old enough to talk or understand words, but I was old enough to know something was wrong.

I recently spent five days in a photography gallery at the Cleveland Museum of Art looking at a collection of 47 photographs taken by Frank Gohlke (Image 18) and Emmet Gowin (Image 19) over stretches of time in the aftermath of the Mount St. Helens eruption. The curator, Barbara Tannenbaum, had brought together these two artists who had worked individually to photograph the aftermath of the 1980 eruption, and she had arranged them in a way that captured the confusion I experienced as a child. Originally, I set about analyzing the exhibit to figure out why it worked as such a visceral memory-evoking experience, but what I found is interesting because it is not just a case study of an exhibit playing with my memory. Rather, this exhibit leveraged people's general perceptual and image-schematic experiences to evoke responses appropriate for the scale and scope of the disaster.

Using Oscillation to Create a Sense of Instability

The curator created a sense of disorienting instability by arranging the images in particular sequences, sequences which oscillated

Image 18. Frank Gohlke (American, b. 1942), *Inside Mount St. Helens Crater, Base of Lava Dome on the Left* (detail), 1983. Gelatin silver print; 45.3 × 55.9 cm. The Cleveland Museum of Art (gift of museum members in 1989).

Image 19. Emmet Gowin (American, b. 1941), *Spirit Lake and Mount St. Helens, Washington*, 1983. Gelatin silver print; 11 x 14 in. © Emmet and Edith Gowin; courtesy of Pace/MacGill Gallery, New York.

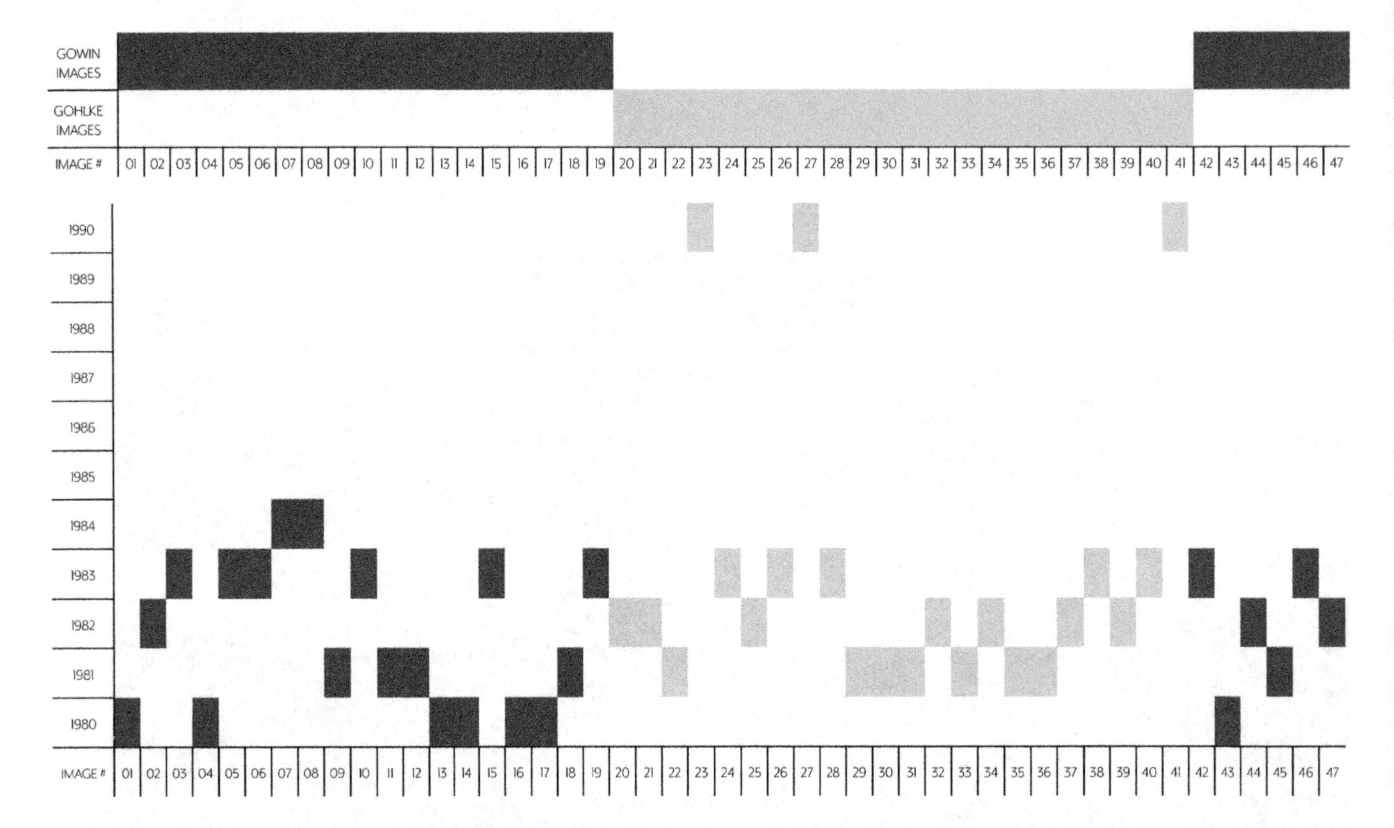

between a set of dimensions including: authorship (who took the picture?), chronology (when was it taken?), viewpoint (was it primarily spectator or participant oriented?), subject (what precisely was in focus in the image), and composition (how the images were internally structured). Systematic oscillation between these dimensions produced instability that gave the exhibit a sense of constant dynamism in the overall narrative of the exhibition.

What became evident as I studied this exhibit was that the arrangement of works was key in highlighting the volcano as a force in creating an unstable scene. Inside this gallery, the volcano became an actor in a story that confused and disoriented its audience through the series of oscillations, and this became apparent to me when I collected data on the content and composition of the images and how those images were arranged in the gallery space.

An Unstable Scene

The Mount St. Helens exhibit presents a non-chronological arrangement of images, and the images were not grouped into categories in the way that a natural history exhibit might have done (e.g., images of eruption, images of flowing lava, etc.). Instead, the images were arranged in a loose sequence that highlighted the instability and unpredictability of the volcano. There were occasional clusters of images arranged around comparisons (of factors like time, location, and scale). What this produced was a storyline of uncertainty and continual destruction. The curator captured the disorienting effect of this instability by returning spectators to familiar scenes in an almost rhythmic sequence. The instability is evident in the constant oscillation between calm and chaos.

Table 7. Authorship: Emmet Gowin (EG) vs. Frank Gohlke (FG).
Table 8. Chronology: 1980–1990.

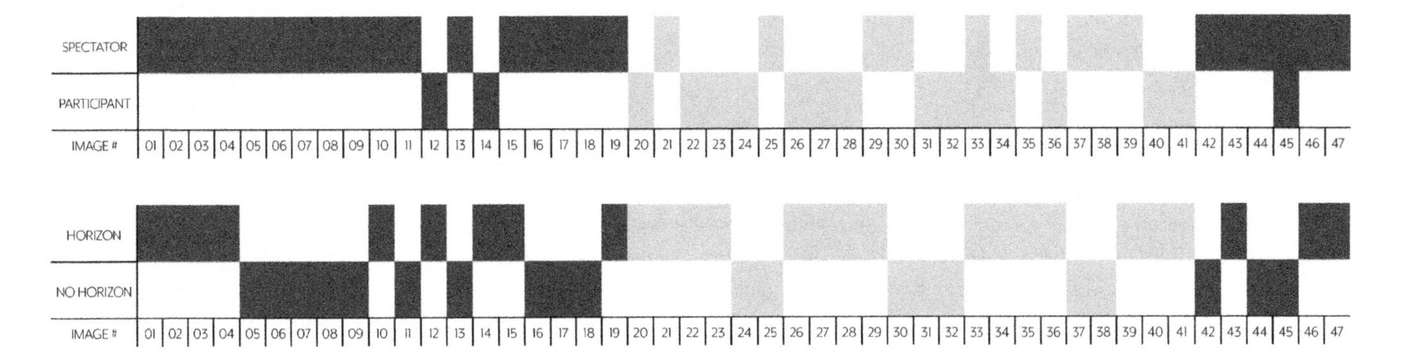

The clusters of categorically related images return the viewer to the awareness that the exhibit is curated and that there is intentionality behind the presentation of images. These clusters act as points of departure from the continual sequence of instability, and also serve functions like advancing the narrative, zooming in, zooming out, and providing scope by means of windowing locations that are scattered near and far from the volcano's crater.

Methods

Several layers of analysis were used to categorize each of the 47 images. These layers included the spacing and grouping of images, the chronological order of the images, how images were sequenced in the exhibit space, image titles, the direction of view within the images, the composition techniques, the use of the horizon line, the use of path and path shape, viewpoint, the use of image features to orient or disorient the viewer, the use of exhibit arrangement as a tool for orienting or disorienting viewers, didactics and maps, the use of figure-ground organization in the images, the use of the schematic directionality of implied motion or terrain structure in the images, the sequencing of images shot from air or ground, the presence or absence of the volcano in the frame, and finally, the agency of volcano (quiet, active, trace activity, or unknown agency). Following this data collection, informal interviews with the curator were conducted to confirm and refine the analysis.

Overall Description of the Exhibit

Approaching the gallery, the viewer encounters a floor-to-ceiling banner with a representational image of Mount St. Helens and the title of the exhibit (other text is visible, but not legible from this distance). Walking into the exhibit, the viewer turns to the left to see the first image in the exhibit.

The viewer moves from a summary view of the wounded volcano (1), to two photos that describe the slope of the crater (2, 3), and then a view of the valley, as if to set the stage for the story (4). Here is the main character, and here is the setting. The next four images are near views of debris flow and rivers (5, 6, 7, 8), followed by three images of veining braids of streams and debris flow (9, 10, 11), and then three images of different scope illustrating the massive treefall caused by the blast—first, a view of the side of a hill covered with dead trees (12), second, a sweeping view of a valley with treefall fanning out in a patterned trace of the direction and movement of the blast (13), and third, a near view of dead timber and stumps in an otherwise desolate landscape (14). These are all aerial images that are characterized by a distinctly spectator-focused viewpoint.

At this point, the viewer encounters the main exhibit banner as seen from the doorway. The viewer is now close enough to the banner to read the description of the exhibit and to view a CGI video simulation of the 1980 eruption compiled with archival images from the National Park System. This looped video gives a sense of the massiveness of the eruption and provides context for viewing the images in the show.

Table 9. Viewpoint: Spectator (S) vs. Participant (P).
Table 10. Horizon Line: Present (Yes) or Absent (No).

Moving past the banner, the viewer reaches a stretch of photographs depicting the devastation of Spirit Lake (15, 16, 17, 18, 19), and then the exhibit takes a turn: up until this point the exhibit has been seen largely from the spectator viewpoint (with the exception of images 12 & 14), but now the sequence of photographs begins to include strong concentrations of participant viewpoint. This new perspective is from the viewpoint of a person walking toward the crater (20, 21, 22, 23). With this change in viewpoint, the images have become larger and continue to increase in dimensional size until the viewer is standing inside the crater at the foot of the lava dome (24). This image of the crater floor is the most palpable image from the participant viewpoint, and immediately to the right of this image is a massive spectator view of the rim of the crater that feels like a hybrid of participant and spectator viewpoint (25). This juxtaposition of the participant and spectator view of the crater creates a dynamic sense of salience for the crater as topic.

The exhibit now begins to concentrate on the deforestation wrought in the aftermath of the eruption—a mix of comparison images (26, 27) and individual images showing dead standing trees, living trees, fallen trees (28), ironic patches of clear-cut forest from logging operations amid forest blow-down (29), a lush forest (30), and a freshly flattened forest (31). The visual scope pulses by providing contextual wide-angle views and then contracts to close-up views of broken trunks scattered in a ravine (32), a scenic overview of a logging area (33), and downed and standing timber covering hillsides (34).

The exhibit crosses again into a view of the volcano (35) and begins to explore concepts of flow while the viewer is presented with images of a river (36), flows of pyroclastic ash swamping huge swaths of land (37, 38, 39), before and after images of scrubby regrowth on ash covered ground (40, 41), and views of dense, wet ash blanketing the landscape under its massive weight (42, 43, 44, 45). The exhibit wraps up with two final views of the volcano (46, 47), the last one being of the intact back side of the volcano, which looks serene and peaceful (47).

Charting the Sequencing of Images

The following set of tables help to visualize the pattern of sequencing and oscillation at play in the exhibit. The image numbers are on the horizontal axis (numbered 01–47), and the dimension in focus is on the vertical axis. In all of the tables, the photographer is indicated by grade of color. The tables each look at one dimension: authorship, date, viewpoint, and whether or not there is a horizon line in the image.

Table 7. Authorship: Emmet Gowin (EG) vs. Frank Gohlke (FG)

Immediately it is clear that the exhibit is divided up into three segments: the initial segment features the photographs by Emmet Gowin (dark grey), the next segment features photographs by Frank Gohlke (light grey), and the final segment returns to featuring photographs by Emmet Gowin (dark grey). There is the beginning of a rhythm here, and it will be interesting to see how this plays out along other dimensions.

Table 8. Chronology (1980 - 1990)

The images were not presented in chronological order, but bounced around over a ten-year time period in a seemingly sporadic manner. There are pairs of images (and one triplet) that occur at the same time, but there are more occurrences of year-switching from one image to the next.

Table 9. Viewpoint: Spectator (S) vs Participant (P)

The most oscillation between viewpoints occurred within the set of Gohlke images while the sets of Gowin images were more stable in terms of viewpoint. It is almost as if there are three segments to this story: a long flight around the area (initial Gowin images), landing and walking around the volcano (Gohlke images), and getting back into the plane to get another comprehensive view before leaving the area (the final Gowin images). In effect, viewpoint oscillations in this exhibit provided a strong contextual background and told a story. Sometimes an image contained both viewpoints, and this is indicated in the table with both the S box and the P box being colored (cf. 33 and 45).

In terms of hierarchy, we could say that there is the **exhibit** which is made up of **image groups** which are made up of **images**. At each level of this hierarchy, we can apply the rule of thirds to see what is salient for that level. This decomposition of the hierarchy also establishes the relationship that each level has to the level above it. If we look at the images in three groups: Group 1 (01–19 by Gowin), Group 2 (20–41 by Gohlke), Group 3 (42–47 by Gowin), it is clear that Group 2 is the salient group at the exhibit level because this group of images focuses most on the volcano from a participant viewpoint, and Group 1 and 3 are background information. At the image group level, the Group 1 and 3 each have bleeps of salience (where the viewpoint switches in 12–14 and 45), while the Gohlke image group is marked by not having any discernible salience (which I would argue is rhetorical). It might be fair to say that the oscillation in this group of images is not principled by the viewpoint variable, and that something else is structuring the oscillation (perhaps some other element of content). Finally, at the image level (not visible in these tables), each image can be judged by applying the rule of thirds grid onto the image and using the rule of thirds in the traditional sense.

The Gohlke images provided a strong part-to-whole relationship with their more frequent switching back and forth between viewpoints. Groups 1 and 3 have strong internal cohesion as stable image groupings where there is little change except for their own bleeps of salience. Group 2 has strong internal cohesion as an unstable image grouping where there is continual change.

Table 10. Horizon Line: Present (Yes) or Absent (No)

There is fairly rhythmic alternation between images with a horizon line and those without a horizon. It is as if the horizon line orients the viewer by showing them where they are in the contextual scene before pushing in (zooming in) to focus on an area within that contextual scene. There is a switching back and forth much like the way that people who are coordinating tasks will switch their attention back and forth between the two tasks. The switching is so frequent that the switch becomes an indicator that the change is meaningful. The switching is between two types of images and the rhythmic switching relates the first type of image to the second type as a connecting device in the visual story. In terms of horizon lines, the Gohlke images act as a kind of visual palindrome, having a patterned ordering that might even be chiastic.

Conclusions Drawn from These Tables

These tables show how the curator told the story by breaking the exhibit down into several slices or views, enabling something new to be gleaned from the data in isolation. First, the curator was able to create an unstable scene by sequencing the images in ways that built a multi-leveled story. Second, the curator built dynamism into the exhibit by making the most of participant and spectator viewpoints. Third, the curator established the image of the volcano as a topical agent in the story by using certain types of images of the volcano as points of departure into new chapters of the story.

The Literary Qualities of the Exhibit: Volcano as Agent and Topic Moderator

A recurring image in this exhibit is the scenic view of the volcano (it occurs in eleven times: 1, 10, 15, 19, 20, 22, 23, 35, 39, 46, and 47). Images 24 and 25 are dramatic views of the volcano mixing both spectator and participant viewpoint together in each image. The volcano image is returned to at several intervals, acting almost as a topic moderator to relate one topic to the next, to turn to a new topic, or to return to the topic of the volcano itself. The image of the volcano helps refocus the viewer in the oscillation between images of the volcano and images of the destruction caused by the volcano.

Moving through the exhibit from the first to last image, the story forays into the devastation in the surrounding landscape, returning to the image of the volcano and its crater to remind the viewer of the source of the destruction. The exhibit begins and ends with the image of the volcano—the first image shows the "front" of the volcano, with the gaping hole blown away during the eruption, and the final image is of the "back" of the volcano, presumably resembling its symmetrical and undisturbed pre-1980 morphology. This book-ending of the exhibit with images of the volcano almost seems literary and acts to keep the salience on the topic: Mt. St. Helens (the actual volcano itself), rather than on the acts of destruction created by the volcano's act of eruption. In this way the volcano becomes a character that has active agency in a story.

Other aspects of the exhibit contribute to this literary sense. For instance, throughout the exhibit, the viewer is taken off into the surrounding wilderness to see scenes of destruction and then the viewer is returned to the crater as if to maintain the topicality of the volcano in the discourse structure. It has the biographical qualities of an adventure narrative, and these forays to scenes of destruction serve as character development throughout this exhibit-story.

Perhaps this could be called an out-and-back approach that keeps the volcano as the primary topic, where the viewer encounters a scenic view of the volcano or a reorienting view of the volcano throughout the exhibit. The images used in reorienting the viewer on the volcano share similarities that are worth exploring: first, they all contain path elements and second, those path elements connect two salient points (or *figures*) in the composition. An out-and-back approach makes sense considering that the flow of information in the overall exhibit is an oscillation, a kind of back-and-forth. And the fact that the composition of the images conveys a path shape for the eye to follow out-and-back only reinforces that oscillation is a rhetorical strategy (both at exhibit level, as well as internal to the composition of images).

I intentionally did not read the description of the exhibit on the main banner (didactic) until the end of my analysis several weeks later. I wanted to see how the exhibit reflected the curator's intent and wanted to see if the exhibit followed the "show, don't tell" rule of thumb. After drafting initial findings from my notes I returned to read the banner. Below are two paragraphs taken from the banner (my emphasis):

> For months after the eruption, the only access to the mountain was to fly over it. Mount St. Helens was Gowin's first experience photographing from a plane; Gohlke had shot from the air for one previous project. **Gowin, who went on to work extensively with aerial views, said that "seeing that landscape for the first time from the air was a revelation." Aerial photography extends human vision to offer what seems like a divine or universal, rather than personal, perspective, evoking in the viewer a new relationship to the landscape.**
>
> **Both artists explored dizzying downward angles where the image fills the entire composition, denying the viewer any horizon line to separate down from up.** Many of Gowin's images first read as abstract patterns, but this was not intentional. "What may look extremely abstract to someone else may look extremely descriptive to me," he said. The newborn landscape, whether seen in macrocosm from the air or microcosm on the ground, reveals the tracks of the immensely powerful geological and climatological forces that continue to transform it.

This is important because it shows that simple perceptual elements in the photographs, and the arrangement and oscillation of differing types of images, achieved a thematic effect that mimicked the effects that the artists themselves felt during their creative process of "documentation," or the production of the work itself.

Discussion

On one level, the exhibit puts the viewers into the minds of the two artists and creates intense emotional prompts for patrons who visit the exhibit with varying degrees of interest, such that a casual passerby might notice the destruction of the volcano and someone who spends more time looking at the exhibit as a whole understands the sustained destruction. Both the casual viewer and the dedicated viewer experience a sense of disorientation, but it is not the same disorientation—the casual viewer may not perceive the decade-long time span and still feel that the destruction portrayed creates a sense of disorientation, while the dedicated viewer experiences disorientation over the decade of destruction.

These effects were achieved through the careful organization and presentation of work that other artists did. This is the heart of curation: being able to relate details in order to weave some sort of narrative about the collection and creating a viewing window into a story of some other world of experience. The curator's use of the oscillation of viewpoint, agency, subject, path shape, and other features created a story of disorientation and sustained destruction in a collection of 47 black-and-white photographs.

At this point, it should be clear that the use of oscillation evokes emotional and cognitive effects in engineered experiences (like curated exhibits in this example), and that oscillation is a useful tool for shaping people's experience of the world.

Oscillation as a Tool for Experience Design

What can be taken away from this case study that makes experience design easier? Oscillation can be used to effectively create disorientation, sustain attention by requiring vigilance to detail and by effectively creating figure-ground comparisons between different features of experience, and it can work to tell a story that both distances the viewer from and also immerses them in details. **In other words, oscillation gives and takes away clarity as a rhetorical tool for advancement on the story line.**

You can use oscillation with any kind of stimuli, and it doesn't need to be pictures in frames on walls. Oscillation might be changes in light or temperature, or sonic variation in drone tones, or in levels of visual detail and access privileges to information that create participant and spectator roles.

Tying Things Together: Paths & Nodes

6

> A path is created in a floor when an independent pattern emerges as a figure against the background of the rest of the floor.
> —Thomas Thiis-Evensen

At this point, we've seen paths deployed in numerous ways as armatures for the designed experience (Chapter 3). Paths structure experience and they help us see experience as a story that unfolds (Tool #6 and Chapter 5). We can think of paths as narrative arcs (Chapter 5). We can think of paths as tools for controlling viewpoint (Tool #3). And we can think of paths as primary places for content with pauses for reflection at the nodes along the path, or just the opposite, where paths connect nodes of content, leaving the moment of reflection to occur during the time spent along the path itself (Chapter 3).

Paths connect elements of the design space and it is therefore worth giving a little more attention to how paths can be designed with intentionality in your installation.

Use a path to lead up to the point of intervention and to lead away from the intervention. In the *BIG MAZE* project (Image 5), visitors spend a long time walking in the maze before they come to the center of the maze. All of the time they spend walking and feeling lost establishes a sense of disorientation, possible confusion,and reinforces the feeling of being lost. This sets visitors up to be surprised by the moment they reach the center of the maze where the path they have walked becomes visible to them from a spectator perspective. Think of this experience as disorientation culminating in clarity. If we think about the maze as a narrative, the rising action is the first portion of walking in the maze and feeling lost, the climax is the moment when everything becomes clear, and the walk out of the maze is kind of like the denouement of the story.

The path designs for remoteness (Figures 10 & 11) have a similar effect. The path leads the walker through the woods, purposefully disorienting the visitor until the confusion of the path opens into a space that serves as a zone of refuge. The visitor can choose to spend as much time in this refuge zone as they want to, and then as they begin to follow the path away from the refuge zone as they walk out of the forest, the landscape opens up to them and they can see the path ahead, bringing orientation back to the person walking on the path.

In IKEA (Chapter 5), the path through the store told a story that could be customized based on customer desires as they wandered through the layout of a multi-roomed home. The views into adjacent rooms and the transitions between rooms, bait the customer to walk into the next room, creating an ad hoc chain of the experience of walking through a home that doesn't really exist, but which could be yours for a certain price.

In the Mount St. Helens photography installation (Chapter 5), the path moved gallery visitors back and forth between participant viewpoint and spectator viewpoint. Over the course of the experience of the exhibit, this oscillation of viewpoint told a story of orientation and disorientation that gave the viewer a sense of the sustained chaos of the scenes conveyed in the framed photographs.

Sensory art that focuses on a particular sensory channel (such as smell, or sound) is often presented as a singular installation focused on the sensory channel under investigation. It is not usually path based. Maki Ueda's *Olfactory Labyrinth* (in Tool #6), on the other hand, turned the installation into a space for inquiry by letting users define the path that they took. It was different than many sensory installations which usually present the intervention at stationary nodes; the very structure of *Olfactory Labyrinth* was path-driven.

Distributing Information across Path Segments and Nodes

You can design paths with nodes that serve as spectator zones and segments that serve as participant zones. You can also design a path where things are reversed—where nodes are moments of deep immersion in some sensory environment, and path segments are moments of clear spectatorship onto the path ahead and, eventually, onto views upon the upcoming immersive node.

A path might connect multiple interventions like beads on a string (as in the IKEA example), or the path might lead to a single intervention (as in *BIG MAZE*), and the path might even be the intervention (as in the *Olfactory Labyrinth*). Looking back to Lynch's primitives of the city (Tool #1), we can break the end-to-end path up into the nodes and segments, and both of these elements can be the places that we insert the intervention of some sensory information or art experience. Sometimes we put interventions at nodes (this works great for spectator viewpoint because nodes often afford views onto something), and sometimes we put interventions in the segments (this works great for participant viewpoint because segments often afford immersion).

Think about the differences between a sight-seeing tour of Paris and a trip to a haunted house. In Paris, you'll bounce around from point to point, and each of the points is an activity node in your experience. In the haunted house, you'll walk through corridors and become frightened along the way, and these corridors are like path segments where the activity is taking place. The Paris trip is node-to-node and the haunted house is segment-to-segment. Note that you can have a segment-to-segment experience of Paris, but often people retell the "highlights" of a trip by describing the activities they did on the trip rather than the movement between activities. A trip through a haunted house has nodes, but those nodes are not nearly as terrifying as the anxiety of not knowing what is coming next as you walk down the hallway toward the next node.

Sometimes you can design a path that makes equal use of the nodes and the segments for different types of interventions. This makes oscillating between participant and spectator easier because you can separate out the segments from the nodes and present one viewpoint in the nodes and the other viewpoint in the connecting segments, such as:

- participant information in the segments, spectator information at the nodes, and
- spectator information in the segments, participant information at the nodes.

You can also combine viewpoints and present them together in blended viewpoints in either the path segments or the nodes in order to focus on some blended content, such as:

- both participant and spectator information in the segments, minimal information in the nodes
- minimal information in the segments, both participant and spectator information at the nodes

Segments can be Participant-Oriented, Spectator-Oriented, or Both

Think about traveling from the country into the city. As you drive along the segment of path toward the city, eventually you start to see the city on the horizon. This is an example of how a path segment can provide spectator viewpoint.

Think about walking through the woods. As you walk along a path segment that is under the cover of the tree canopy, your view is contained in a sort of conduit and the feeling is immersive because it only provides a participant viewpoint. And if the trail bends up ahead, you are unable to take a spectator view of what lies ahead of you.

Sometimes you will follow a path segment and be able to see what is up ahead of you as a spectator while feeling fully immersed as a participant. Think about driving through a tunnel, you experi-

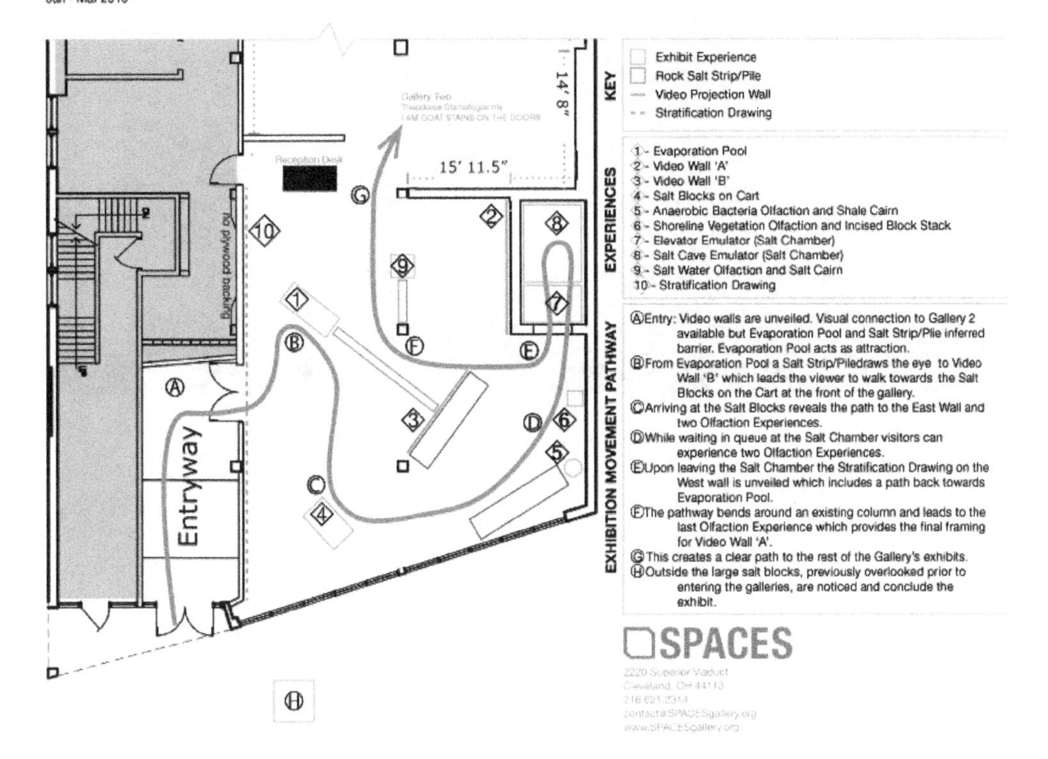

ence the tunnel all around you but you can also see the light at the end of the tunnel. You feel immersed but you remain oriented.

Case Study: Layering Viewpoint Information Along Paths and at Nodes

In 2016, I designed and installed a site-specific project for the Geologic Cognition Society with architect Dru McKeown and sound designer John Daniel, called *UNDERNEATH IS BEFORE*, in Cleveland Ohio at SPACES, a gallery located just up the hill from the entrance to an active salt mine that is almost 2,000 feet under Lake Erie. The installation successfully replicated the feeling of being in the salt mine (a tour of salt mine workers confirmed that they felt like they were inside the mine), but the installation was also focused on attempting to represent the sensory conditions of the moment in geologic deep time when the salt deposit was forming. The installation was immersive and it combined sensory layers to tell a story about the accumulation of salt as the ancient Silurian seas slowly evaporated 300 million years ago. The experience was strongly path-structured (Figure 15), building up layer by layer through a combination of path segments and nodes which used combinations of sensory viewpoints to bring the visitors to the climax of the experience in the quieting hollows of a salt chamber that was reconstructed from salt mined for the installation. The point of salience was on the salt chamber, located three-quarters of the way along the path through the installation.

Figure 15. Floor plan design graphic of path and nodes in installation, *UNDERNEATH IS BEFORE*, by Geologic Cognition Society at SPACES Gallery (2016). Figure by Dru McKeown. Used with permission.

As the visitors moved along the path toward the salt chamber, layers and sequences of information combined sensory viewpoints to accumulate (much like evaporating salt in the ancient sea combined in sedimentary layers), and the path helped build up the experience toward the story climax in the chamber.

The installation itself was structured by a source-path-goal schema which followed a narrative arc, placing the moment of climax at nearly three quarters of the way along that path inside the salt chamber (Figure 15, see Diamond 8). Each individual layer in this installation also had its own internal structure of the source-path-goal image schema, where the visitor started at one node (source), saw or otherwise sensed some goal (goal), and then moved toward that goal by following a predetermined path (path). Upon reaching the goal, the next goal came into sight and the experience pulled visitors through the overall exhibit path in an incremental and segmented series of smaller paths. This was a strategy that enabled the psychological suggestion of a path to be designed into the space without being explicitly marked in the space. The goal portions of each of these source-path-goal segments acted as attractors that enticed people to walk toward them to better observe the object. Upon reaching the object, the next goal came into the visual field.

Here is a point-by-point discussion of the nodes and path segments on the installation graphic (Figure 15).

Layer 1: Visual Overview and Thematic Confrontation

Visitors enter through the front door and walk down a short hallway that opens onto a spectator viewpoint of the gallery (Image 20). At the threshold of the doorway to the gallery (Figure 15, see Circle A), visitors are confronted with two walls of video projection (Figure 15, see Diamonds 2 & 3) of ocean water waves which slowly solidify into static images of banded rock salt (Image 21). The gallery guide describes this moment in the exhibit with "*Active oceans turn to static salt deposits; water evaporates and leaves a crystal trace. This spot where you are standing has seen a lot in the last 300 million years.*"

Layer 2: Entering the Space and Walking toward Salt Rocks

After pausing for Layer 1, visitors become participants by entering the gallery space and feel the space as an immersive whole. Visitors walk to the nearest intervention node, the evaporation pool (Figure 15, see Diamond 1). This pool (not visible in the image) floats in a linear pile of rock salt that juts out from the angled projection wall like a peninsula into the room. The angle of the salt pile psychologically blocks visitors from taking a leftward path through the exhibit (Figure 15, see Circle B) and instead deflects them to the right to move counterclockwise through the exhibit. This blocking is an application of the image schema of **barrier**, and because of the extent of the barrier and its angular orientation, the path logically opens to the right where an object (an industrial cart loaded with 2,000

Image 20. Layer one. *UNDERNEATH IS BEFORE* (2016). Geologic Cognition Society for SPACES. Courtesy of the artists.

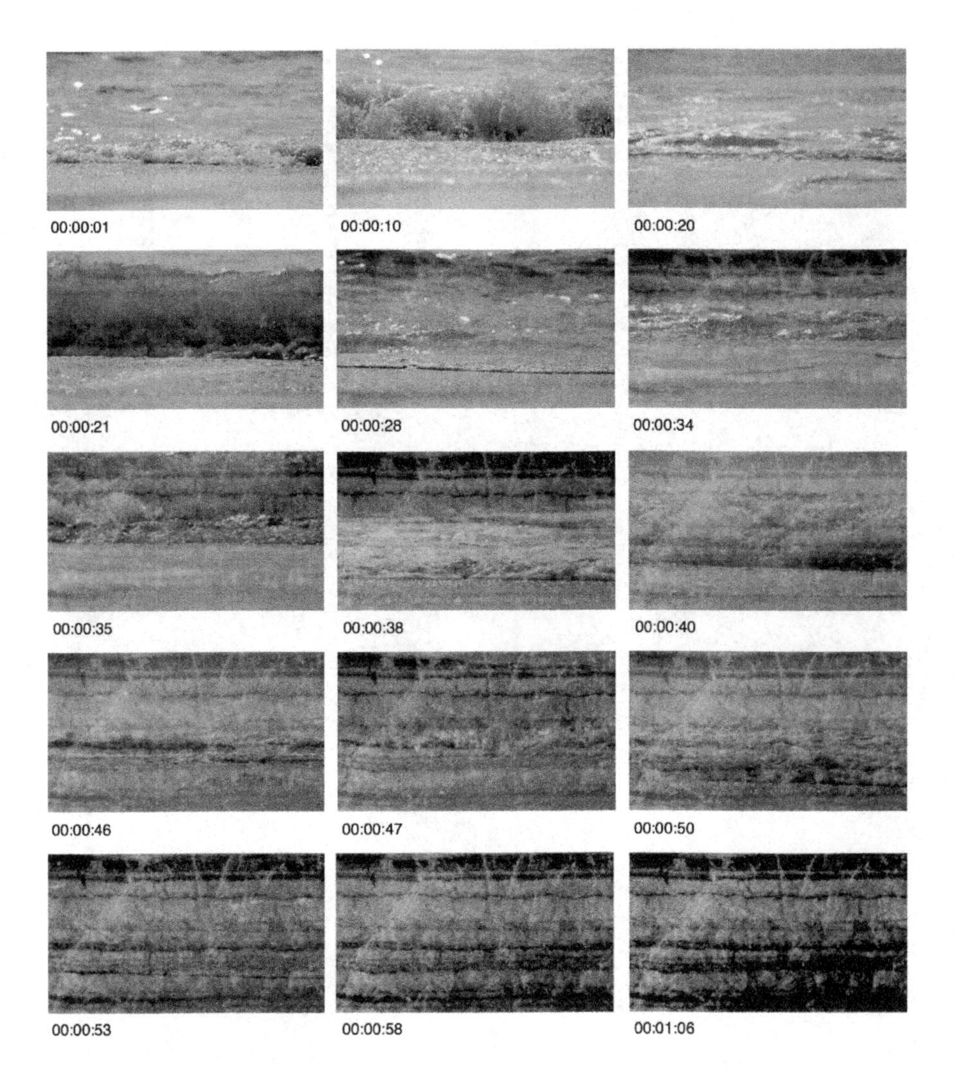

pounds of salt rocks— Image 22) sits isolated in the gallery and acts as an attractor and goal for visitors as they move along the path. The cart (Figure 15, see Circle C) is lit from above with two spotlights while the rest of the gallery corner remains unlit. When visitors spot the cart, their attention is directed to the cart with spectator viewpoint—they observe it from afar and move toward it as a destination.

Layer 3: Sensory Basin and Sensory Corner

Visitors circle around the cart looking at the salt rocks. The rocks are visually interesting and provide a moment of deep looking, visitors are absorbed in a visual and haptic participant experience as they look at and touch the giant salt rocks. Once they have completed walking around the cart, a recess behind the projection wall (Figure 15, see Diamond 3) visually opens up to provide a spectator

Image 21. Stills: Moving from active ocean scene to static stone scene. *UNDERNEATH IS BEFORE* (2016). Geologic Cognition Society for SPACES. Courtesy of the artists.

Image 22. Layer two, visual goal. *UNDERNEATH IS BEFORE* (2016). Geologic Cognition Society for SPACES. Courtesy of the artists.
Image 23. Layer three, visual goal. *UNDERNEATH IS BEFORE* (2016). Geologic Cognition Society for SPACES. Courtesy of the artists.

Image 24. Shale cairn. *UNDERNEATH IS BEFORE* (2016). Geologic Cognition Society for SPACES. Courtesy of the artists.
Image 25. Anaerobic bacteria olfactory station. *UNDERNEATH IS BEFORE* (2016). Geologic Cognition Society for SPACES. Courtesy of the artists.

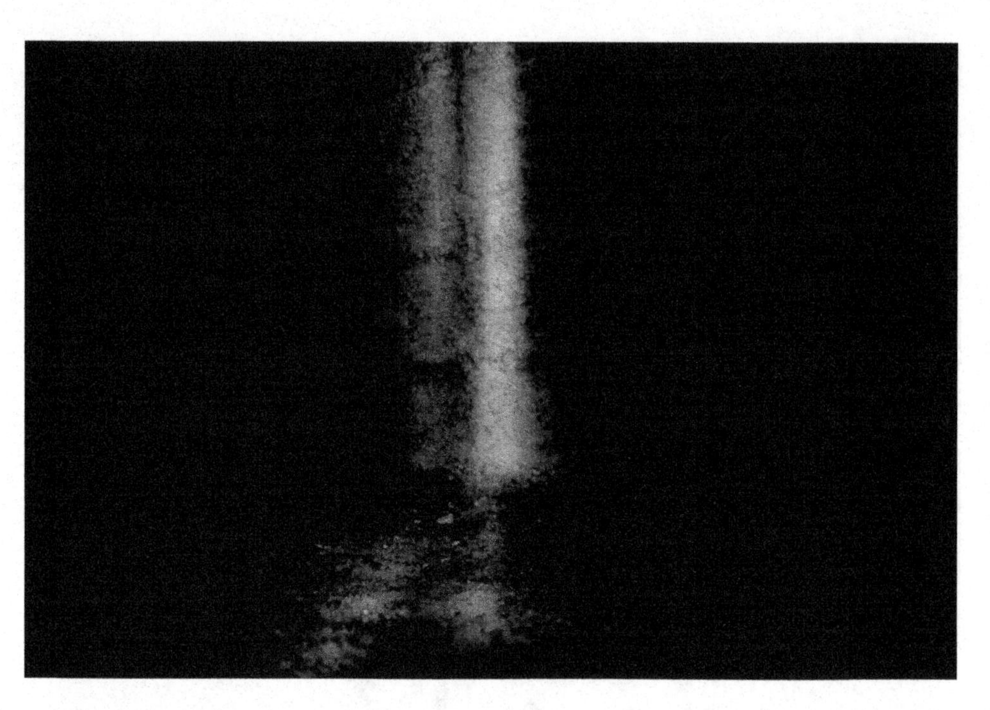

viewpoint of the next participant node: a space with two sculptural objects lying on the floor against the wall (Figure 15, see Diamonds 6 & 5; Image 23).

As visitors walk into the space they are confronted with the sulfurous smell of anaerobic bacteria (an olfactory participant viewpoint) from a sensory object that remains unseen until they approach the shale cairn sitting on the floor (Image 24). The angle of the cairn creates a vertex of a triangle with the other elements of the sensory corner and as visitors turn their heads to the right, they see a polyethylene bag (Image 25) suspended from the ceiling, full of aquatic plants and anaerobic bacteria. This bag is the source of the stench as the anaerobic bacteria digests the plants and the bag is off-gassing the swampy smell of rotten eggs. The stench fully immerses the visitor in an olfactory participant viewpoint. The bag drips a slow drip down into a bucket below, which completes the triangle and helps to diffuse the scent. Shale is fossilized anaerobic bacteria, and this node depicted the sensory conditions of a band of shale that released hydrogen sulfide in the actual mineshaft.

This piece recreates that smell while pointing to the ontological relationship between shale and bacteria, and it also creates an immersive participant experience of an olfactory band that visitors walk through on their way to the next goal. While this olfactory corner (Figure 15, see Diamond 5) gave an immersed participant viewpoint of the smell of anaerobic bacteria, it did not give a visual spectator viewpoint unless you followed the scent into a hidden corner to see the source of the scent, and that was when the cairn, bag, and bucket completed the ontological triangle suggested in the form of the cairn and the tension of the hanging bags.

Immediately to the left of the shale cairn are two towers of salt blocks with a matrix of linear cuts carving the blocks in chaotically overlapping transects. Behind this piece is an atomizer diffusing essential oils of sea vegetables into the air (seaweed, sea buckthorn berries, etc.) to evoke a more oceanic feeling and to provide a pleasant scent as visitors approach the climax of the installation behind a curtained doorway (Figure 15, see Circle E), the threshold of the climax.

Image 26. Interior of the salt chamber. *UNDERNEATH IS BEFORE* (2016). Geologic Cognition Society for SPACES. Courtesy of the artists.

Layer 4: The Elevator Shaft

Visitors walk into the simulated elevator shaft (Figure 15, see Diamond 7), a space which is simply curtained off with layers of black fabric that envelop the visitors in darkness, creating a visual and spatial sense of participant viewpoint through the blocking of light. This antechamber has the image schematic structure of containment, and from the beginning of their experience in the gallery, visitors have moved into increasingly smaller spaces one-after-another, in a gradual nesting of space. A soundscape begins to play a sequence of mechanical sounds—drops of water, rattles of chains, and the rushing of air—to mimic the sound of descent in the elevator shaft of the salt mine. Binaural beats and sequenced sounds create the impression of downward movement as sound seems to rush upwards past the visitors in the room. When this sonic piece culminates, it transitions into an ambient piece scored for the salt chamber and visitors pass through the curtains and enter the salt chamber (Figure 15, see Diamond 8).

Layer 5: The Salt Chamber (Climax)

Inside the salt chamber (Image 26; Figure 15, see Diamond 8), the ambient loop sets a tone for the experience, but it does not overpower the sound of movement in the space. 500 pounds of rock salt line the floor of the chamber to create a crunching sound underfoot as visitors walk around and explore the chamber. The temperature inside the chamber is colder than the gallery, but close to the temperature inside the mine. The walls of the chamber are lit obliquely from slits along the back wall with strips of dim, floor-to-ceiling LED lights, casting a slice of light against the 500 pounds of salt that form the walls of the chamber. The ceiling is low to give a feeling of spatial compression and to suggest the weight of the earth above the mine, but the room feels more open after visitors exit the confinement of the elevator shaft. This salt chamber has a container schema, and visitors are inside the container, confronted with solitude, enclosed in a space that feels totally different from the outside gallery space and insulated by the buffer space in the elevator shaft antechamber. The faint ambient sound is coordinated with the *dimness* of the lights: the sound is *low,* the lights are *low,* the ceiling is

Image 27. Sensory basin and video walk. *UNDERNEATH IS BEFORE* (2016). Geologic Cognition Society for SPACES. Courtesy of the artists.

low, the temperature is *low*, and the time spent in the antechamber elevator shaft created the impression of *descent*. All of this contributes to making the salt chamber feel as if it is *below* ground and it is all coordinated by a cross-sensory metaphor of *descent*. It feels otherworldly. This room is the most palpably immersive element in the installation, creating an overwhelming participant viewpoint. As visitors walk around inside the salt chamber, the salt underfoot sounds gravelly, a contrast to the echoing sound of footfalls on the wooden floor of the gallery. The enclosure of the 8'×6'×8' chamber contains sound and heightens attention to the sound as other noises are deadened. As visitors listen to their feet walking on the salt they experience a moment of heightened awareness. Visitors exit.

Layer 6: The Stratification Drawing / Graphite Mural

Upon exiting the salt chamber, visitors move into a more open space that is brightly lit in contrast to the dim chamber they have just exited. As they turn right out of the chamber, visitors are now facing the opposite wall of the gallery for the first time (Figure 15, see Circle E). From the left/front corner of the gallery to the right/back corner, there is a floor-to-ceiling graphite drawing of striation marks like the striations in the salt rocks and salt chamber (Figure 15, see Diamond 10), reinforcing the layered nature of the deposit of salt. The lines of graphite slope downward from left-to-right, creating a kind of slant that points like an arrow toward the door to the next room in the gallery (Figure 15, see Circle G). At room-sized scale with spectator viewpoint, the angular motion of the graphite lines gives a sense of dynamism and pulls the visitor through the next layer and toward the next participant node, another sensory basin (Figure 15, see Diamond 9).

Layer 7: The Sensory Basin and Video Walk

As visitors approach this final sensory basin (Figure 15, see Diamond 9), the space is filled with the smell of ocean water created with a diluted dimethyl sulfide solution that smells like washed-up seaweed and other shore detritus (Image 27).

As they walk past the sensory basin, the projector washes them with the video of ocean waves (Figure 15, see Diamond 2; Image 28) and visitors exit the front gallery and move on to see the other

Image 28. Projection wall. *UNDERNEATH IS BEFORE* (2016). Geologic Cognition Society for SPACES. Courtesy of the artists.

exhibits in the adjacent rooms (Figure 15, see Circle G). But their path encounter with salt is not yet over.

Layer 8: The Salt Boulder on the Plinth Outside

Upon exiting the gallery, visitors pass by a 2,000-pound salt boulder sitting atop a sandstone plinth (Image 29) and decaying in the natural elements (Figure 15, see Circle H). Upon arriving to the gallery at the beginning of the visit, this rock might not have had the meaning that it now has for visitors after they have experienced the salt chamber and other exhibit elements. Some visitors lick the boulder, which sits almost at shoulder level. Others have their picture taken next to the boulder. Still other visitors fail to notice it as they leave the gallery. Over time during the exhibit, the salt boulder delaminates at the seams of volcanic ash as rainfall saturates the banded striations in the salt. Massive chunks of the boulder fall to the ground as each layer slowly separates like calving glaciers, and the layers lay on the ground around the plinth as a reminder of the layered nature of geologic deep time, reinforcing the notion that time is layered and underneath really is before.

Image 29. Salt boulder on plinth, after Brinsley Tyrrell's (1996 & 2000) *Salt of the City* in the same location. *UNDERNEATH IS BEFORE* (2016). Geologic Cognition Society for SPACES. Courtesy of the artists.

7

Documentation for Planning, Archiving, and Reproducing

Complexity overwhelms people. Minimizing complexity may not be possible as you plan out your experience, but it might be possible to bring clarity to that complexity by creating a series of diagrams of the experience. These diagrams can serve a dual purpose: as planning and operational guides, and also as archival documentation of the work.

Experience-based works often stand alone as self-contained worlds. These works are often internally structured with logics devised for maintaining self-consistency. This rigor is what allows a work to operate as a system and it helps develop an ecological structure for the work. Because of this complexity, new models of documentation are needed for experience-based works.

Michael Mansfield addressed the notion of needing new documentation models in an interview at the Library of Congress. Consider this answer he gave when asked about the models needed for conservation of new media works:

> I'd like to find interesting ways to document the lifecycle of media artworks. This might be out of left field a bit, but artworks like this seem to live and breathe in ways that are unique in the arts and unique in their time or historical place. They grow, or shrink. They respond to their surroundings. They physically evolve. They consume. They age. They die ....In some cases they reproduce. Outside of the box, I think we might benefit from some creative, comparative research with animal sciences, through their documentation of life cycles. We can look at the tools used by zoos and their conservation practices with living specimens. How do they document natural behaviors of a living creature? Perhaps this might generate some new ideas for handling something like an artwork, something that is uniquely human. (Mansfield 2013)

While this is a good start, because art is a human practice, the documentation approach needs to be influenced by more than just the animal sciences. It also needs to be informed by anthropological approaches to mapping cultural complexity.

This chapter focuses on how to break down an experience into partial networks of detail that can be organized as layers of data and mapped out in a series of diagrams. Conceptual tools from archeology and data management tools from spatial modeling help explain the ideas of modeling networks of activities in complex engineered experiences.

Remember that worlds do not need to be complete in order for an audience to find the story believable. This partial structuring of the world comes from an idea and archeological method called *entanglement,* which looks at how relationships structure the flows of matter, energy, and information (Hodder 2012, 105). If a story is thought of as a series of flows of information, and a physical experience of that story involves the flow of matter and energy and information on the part of the audience, then perhaps entanglement can help make sense out of the tangled ball of details in your engineered experience.

Hodder (2012) models the partial networks of flows in archeological data using something he calls tanglegrams, which map the flows of information, energy, and matter in a tangle of complex messiness that preserves the relationships of dependence that exist between the different view layers of the data. The planning and documentation layers discussed in this chapter are really a form of a tanglegram that can be plotted in a database, in a schematic sketch and simulation program, with timeline creation tools, or in a geographic information system (GIS).

Hodder views entanglements as dynamic, which makes them ideal for modeling immersive and responsive activities that audience members experience over specified periods of time. He views entanglements as "*provisional, worked out in practice, temporary and partial,*" full of "*messiness and contingency*" that is uncontained and "*difficult to predict because of the strands that seem to spread out everywhere*" (Hodder 2012, 110). This sounds a lot like the structure of an engineered experience. Fortunately, entanglements are *partial* structures made up of simple sets of relationships linking humans and things together (Hodder 2012, 105), and not everything has to be modeled—only what is relevant to the idea in focus. If we can dissect an experience into the relationships that govern the flow of information, matter, and energy in an experience, then we will be on our way to building a playbook for running the experience. This comes from a simple breaking down of the world, much like Spradley's descriptive question matrix (Table 1), into the categories of different elements of an experience so that each category can be looked at on its own or in relation to other categories.

Entanglements offer a selected view, hence their partial nature. Entanglements don't show everything, only the elements caught up in the ecology of whatever relationship the entanglement represents. This is another way of saying that entanglements are models of some selected set of relations between entities.

Entanglements are nested and composed of lower-level entanglements. You can ostensibly make a master tanglegram that includes every other tanglegram, but that might be overkill, if not impossible. Tanglegrams are diagrams which list the components and concepts in an ecosystem and specify the linkages between those components and concepts. Some components are linked to multiple concepts, and some relationships are unidirectional, while others are bi-directional and multi-directional. It's called a tanglegram because it looks like a tangle of arrows knotted up around concepts and the linked components, and it looks messy, but it summarizes the ecology by showing how it relates to itself.

A similar process happens in spatial analysis with geographic information systems (GIS) that are simple thematic layers of data about a given location. The analytic part comes by forming questions about that location by comparing different data layers. This gives you insight into conditions in that location and helps create causal "*what if?*" scenarios to see how changes to the location in one layer might affect conditions in another layer.

Without tying this practice to any one particular technological solution, various online mapping tools can help create layered views of your experience world, as can simple layers in common graphic design tools. If all of that is too complicated for you, the same information can be mapped out in sketches made of transparency sheets or translucent vellum and then comparing layers simply becomes a process of selecting which transparency sheets to compare and laying them over the map of the physical space for the experience and observing the impact one layer of data has on the other layers of data.

What To Map

As noted in Chapter 1, Spradley's descriptive question matrix (Table 1) helps break down an experience into a series of domains: Space, Object, Act, Activity, Event, Time, Actor, Goal, and Feeling. We can take these domains and derive a series of layers to the documentation of an installation. To show how this covers the semantics of an event, we can correlate Spradley's domains to the six basic questions words of journalism (who, what, when, where, why, and how):

— space: *where*
— object: *what*

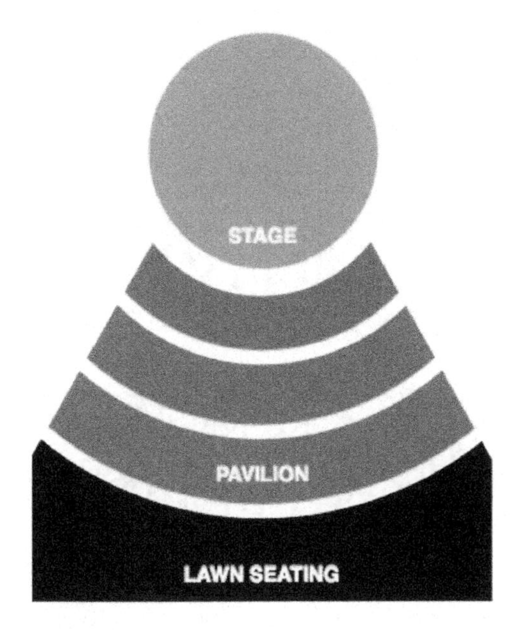

— act: *what*
— activity: *what*
— event: *what, when*
— time: *when*
— actor: *who*
— goal: *why*
— feeling: *what, how*

Spradley's domains appear to provide thorough coverage of all of the structural information in the event plan because using Spradley's domains answers all of the six basic question words. But when working with immersive and transformational experiences, it can be useful to view the world in finer granularity and from different angles that aren't addressed as cultural domains which do not provide all of the categories necessary for comparing some elements of engineered experiences as described in this book. For instance, categories such as attention, viewpoint, emotional triggers, plots, directions of flow (for time, energy, information or matter), ambient environmental changes, or feedback loops are unaddressed by cultural domains. Since these categories are crucial elements of an effects-based approach to immersive and transformational experiences, it is important to map these extra categories as layers in addition to mapping the domain layers. Extending this original matrix is vital for creating thorough and useful documentation for your work. In the remainder of this chapter, we'll look at how to layer the domains and the extra categories of the experience.

The Base Layer: A Plan View Spatial Layout

Create a layer for each of the categories that you use to structure your experience and plot out the relationships, flows, and changes that will occur in the performance/experience. Start by producing a plan view of the event space(s). A plan view is the basic bird's eye view layout of the space. Draw the floor plan of the space and use this as a template for each of the successive layers. This becomes the base space layer on which other layers can be placed to see how they unfold in the event space. When you create a new layer that focuses on a particular topic or theme, use the floor plan template and then add all of the information that is relevant to the topic or theme onto that floor plan. Do this each time you create a new layer and switch to a new theme.

The base layer in any documenting project should define the space that is being used in the designed experience. This is the map of the room, gallery, or venue, presented in a conventional way—demarcating the boundaries and walls, entrances and exits, windows, and other structural and architectural features of the space. If it is an outdoor venue, then provide an outline map of the space, and if the terrain of the site is part of the experience or the design, then provide a contour map of the area. (Figure 16 is a simple spatial layout of an outdoor concert venue.)

After creating the plan view of the space, work to create other relevant layers. Choose from the following layers, some of which may have more relevance to your project than others.

The Activity Layer

In this layer, map all of the activities that will take place and provide annotations with the exact location of the activity, the timing of the activity, and the sequence or order of activities that take place in the

Figure 16. Spatial Layout of an Outdoor Concert Venue.

space. This might vary based on narrative structure, so compare this layer with the narrative layer to identify the order of activities from the narrator's god's-eye view and the character's participant view. In a non-linear narrative, the order of the activities might not line up with their sequence from the god's-eye view.

The Actor Layer

Color-code all of the actors or categories of actors (this includes the categories for the audience). Use this layer to map out each category's planned movement throughout the experience by drawing an arrow. If the actor category changes direction within the space, terminate the arrow and begin a new arrow that goes in the new direction. Arrows for actor movement should only carry a single vector. If you model and plan audience behavior in this way from the beginning, this unique vector rule will help identify points of possible intersection that might be good locations for triggers, plot changes, and audience interaction. By beginning a new arrow with each directional vector, this layer will clarify audience motion and flow.

The Event Layer

Mark all of the events on this layer by indicating the time and space that the event occupies. Create zones that are shaded to allow easy identification of the event sequencing. While other layers will contain similar information, this layer remains relatively uncluttered by only designating the major high-level events that take place during the experience.

The Object Layer (All of the Nouns)

If the audience interacts with the physical environment (such as objects in the space), annotate the locations of these interactions and specify the outcomes of those object interactions.

The Attention Layer

On this layer, draw arrows that represent the direction of the gazes of all participants in the experience (using colors for participant categories), but also mark locations of salience, figures and grounds, and locations of directed attention. Number each location in the sequence in which it occurs. Indicate all of the attention patterns and the change of attention as it is directed during the event. Figure 17 is the attention layer of the outdoor concert venue in Figure 16. The top arrows represent the members of the orchestra and their attentional orientation toward the conductor, represent by the single opposing arrow. The next layer of arrows are pavilion seats with their attention focused toward the stage. The solitary arrow in the center is the soundboard operator, while the opposing arrows in the second layer mark out the security guards monitoring the bottom layer of people sitting on the lawn so as to prevent the lawn ticket holders from entering the pavilion seating area. The arrows indicate the direction people are facing and the direction that they exercise vigilance during the performance.

The Emotional Trigger Layer

Mark the locations and times of the onsets of emotional triggers. This layer will correlate with the temporal map (Figure 18) for a time-based view of the locations for specific triggers. If the emotional triggers are dependent upon each other and need to occur in a crucially ordered sequence, indicate that sequence to show the developmental flow of the emotional fabric of the event. If the audience is supposed to experience different feelings at different points

Figure 17. Attention layer of an Outdoor Concert Venue.

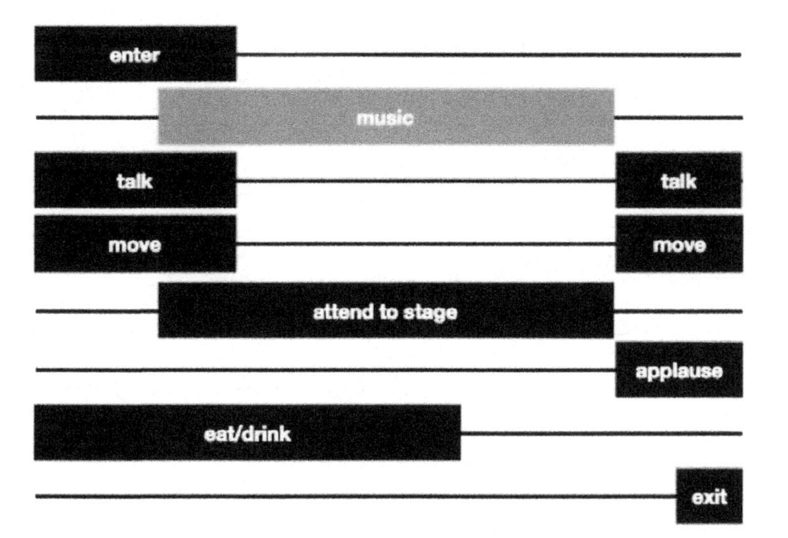

in the experience, mark the areas you want to reserve for those feelings. Note the transitions between feeling zones, and see how adjacent feelings affect adjacent spaces. Do zones modulate audience members' paths differently? If different zones do influence audience member's paths and you designed these zones to influence behavior, then mark the routes between feeling zones with the color-coded participant arrows. If you do not design this into the experience and it ends up emerging during the experience, chart these movements on a participant-by-participant basis and see if the participant categories give insight into why participants traveled through the space along the paths that they did. Also, if your experience has any debriefing element, try to capture simple descriptions of what participants were feeling as they went through the space. Coordinate that qualitative response data with your quantitative audience path behavior to see if patterns emerge as a product of your design.

The Viewpoint Layer

On this layer mark all of the space that uses participant viewpoint, spectator viewpoint, or blends of participant and spectator viewpoint. These spaces will likely be large areas, so perhaps use a shading pattern or texture for each viewpoint designation. Indicate the points of transition where one viewpoint begins to incorporate another viewpoint, or begins to change viewpoint. Oscillations in viewpoint should be marked as well. If different segments of the audience are experiencing different viewpoints, note this with a colored dot inside each of the shaded/textured areas. If the viewpoint depends on the direction of travel through a particular space, do not shade/texture the areas, but indicate with a shaded/textured arrow the direction of travel that correlates with a particular viewpoint. In this way, you may have a participant-shaded arrow going in one direction while a spectator-shaded arrow may go in the opposite direction. Always indicate which audience members experience which viewpoint if there are categories of audience members.

The Actor-Behavior Layer

You may want to map the major activities performed by actors in your documentation. This could be helpful when tracking the roles that confederates play in the experience, or it could be used to simply mark out who does what where (Figure 19).

The Plot Layer (Provide Spatial Coordinates for All Plot Advancement)

If there is a narrative/storyline to the experience, indicate spatially where different elements of the plot take place. For example, if the space is laid out in a winding linear path, indicate with shading which parts of the path are devoted to the exposition, rising action,

Figure 18. Event-Time Flow Model of an Outdoor Concert Performance.

climax, falling action, and denouement (or whichever model of plot structure your experience uses). If the space and plot are coordinated such that in the layered plan view the climax falls on one of the spatial thirds (rule of thirds), when a grid is superimposed on top of the map layer, mark this to make explicit that the point of narrative salience is the climax and that it occurs in that particular spatial zone. It may not be obvious in other documents that there is a blending of physical space and narrative space. If in your designed experience the audience members have an overall spectator view of the experience space (i.e., they actually get to view the space from plan view), then this privileged spectator view plays the role of a visual narrator of sorts, much like the narrator from a story (or a sports broadcaster giving play-by-play calls from a stadium box). By making this narrator/spectator viewpoint more explicit in the documentation, you clarify that there is a spatial plot metaphor that structures the experience. This concept may be difficult to communicate otherwise. If the narrative structure is nontraditional, non-chronological, and/or non-linear, mark or shade the different physical areas in which audience members experience plot-advancing techniques. Create a master numbering system as if the narrative structure were linear and traditional, which will allow you to see the overall plot structure from a god's-eye view (i.e., the perspective of a divine spectator) of the experience space. Also have a sequential lettering system that follows the spatial path that audience members will follow as they move through the space to indicate what the audience members see. Correlate letters and numbers in a list to see which numbered plot element the audience is encountering in which lettered experience sequence.

The Time Flow Layer

If the experience is set up to have some of the audience experience one time model (e.g., ego-moving time) and the other part of the audience experience another time model (e.g., time-moving time), draw long arrows for each of the directions that the audience experiences as the flow of their temporal model. Block off event durations and mark the transitional stages of the event.

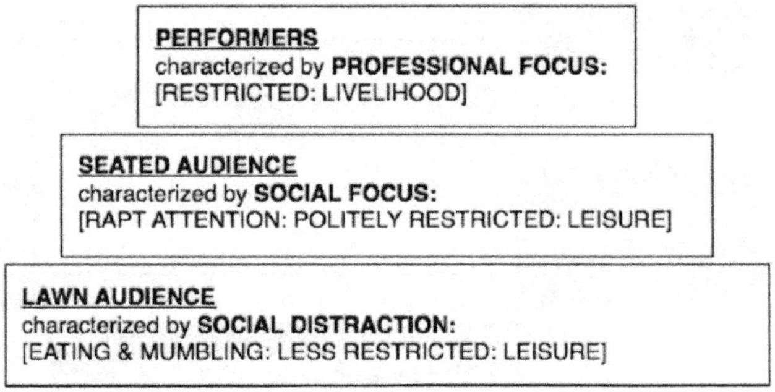

PERFORMERS
characterized by **PROFESSIONAL FOCUS:**
[RESTRICTED: LIVELIHOOD]

SEATED AUDIENCE
characterized by **SOCIAL FOCUS:**
[RAPT ATTENTION: POLITELY RESTRICTED: LEISURE]

LAWN AUDIENCE
characterized by **SOCIAL DISTRACTION:**
[EATING & MUMBLING: LESS RESTRICTED: LEISURE]

Figure 19. Actor Behavior Model of an Outdoor Concert Venue.

For example, an outdoor orchestra performance might have a time flow with overlapping channels that looks something like Figure 18 on the previous page..

The Energy Flow Layer

Chart the movement of audience, support staff, and other people through the experience space. Use color-coded arrows. You may want to use arrows that are stylistically different from the attention layer arrows, in case you want to compare the attention layer and the energy flow layer.

The Information Flow Layer

If the audience gains information throughout the experience, indicate where that information is encountered and connect it to the locations in the experience where that information will be required/used. For example, say an audience member has to learn a passcode at an early stage in the experience and then they have to use the passcode to unlock a new level of experience later in the event—indicate the connection between the reception of information and the later transaction that consumes the information. If visual information moves through the space with the audience members, indicate that they travel in sync with each other, or at whatever rate they move with respect to each other.

The Matter/Object Flow Layer

If objects and matter are moving through the space, indicate the direction of their movement, noting the sources of those objects, the paths they take, and their destinations. Indicate speed and orientation of the objects. Annotate all of the encounters that audience members have with these moving objects.

The Ambient Environmental Change Layer
If there are zones that change with respect to ambient characteristics, mark out those zones and note how they interact with the physically adjacent layers, and also note the evolving context(s) as each adjacent layer changes ambient characteristics. For instance, if Zone A is a cool dark space and Zone B is a warmer dark space in the first sequence, but in the second sequence Zone B becomes a warmer light space, how do all Zone Bs relate to each other, how does Zone A relate to both Zone B sequences, how does second sequence Zone B begin to change Zone A over time, and so on? How does Zone B emerge into its second sequence and how do all physically adjacent zones and temporally adjacent zones drive that emergence? Ambient environments can be stable over long periods of time or they can constantly evolve on fast or slow time scales. Mark the stability, duration, and speed of ambient zones. If there are relationships between ambient zones and information flows in spatial narratives, then having this information about ambient stability will be useful for planning scene changes and other dynamic plot events.

The Feedback Loop (Temporal & Spatial) Layer
All sensors that trigger a feedback loop should be plotted spatially, and if multiple types of feedback loops are being used, the process engine for each loop should be specified, as should the product being fed back into the system as the new input in the spatial and temporal location in the experience space. This can help you keep track of changes to your experience space that you might not have anticipated. If you can simulate this, do so ahead of the experience (even if you merely simulate it by drawing on a piece of paper with a pencil). Because feedback loops are so dynamic and can create massive generative disruption in a system, being able to see how they will work out in your designed space will help you build stronger spatial plot structures and help you predict atmospheric changes in the environment.

The Audience Interaction Layer
If there are points in your experience in which you want the audience to engage each other, mark these locations and use the color-coded participant categories. Indicate all of the differences that you have given to the different audience members. If you have given training and priming activities to different groups in your audience, annotate that training and the likely knowledge, expectation, and/or viewpoint conflicts that the audience members will have in their interaction. Annotate the sites of interaction using Clark's joint activity dimensions (1996) to show if audience interaction is *scripted, formal,* or *cooperative*. Also use this layer to plot out the zones for different joint activities and indicate which portion of the audience are engaging in which joint activity, and which direction they approach the activity from or how much knowledge they are expected to bring to the joint activity. Try to describe as completely as possible the *who, what, when, where, why,* and *how* of audience interaction. Human interaction is always messy, so expect this map layer to be messy as well.

The Audience Decisions Layer
If there is a decision tree or outline of all the possible decisions (like an infographic flow chart) that can be made in the experience, then use this layer to map out all of the points of audience decision-making. Annotate what the range of options are and provide a basic wire-framing for the possible paths taken through the decision tree (if complexity is not an issue). This might not be necessary for planning if you are merely interested in seeing how an audience responds when presented with some set of decision-points. However, in documenting the project, it might be useful for you to have a record of the range of decisions that were actually chosen, and to be able to coordinate this data with the other layers in the dataset.

Image 27. Sensory basin and video walk. *UNDERNEATH IS BEFORE* (2016). Geologic Cognition Society for SPACES. Courtesy of the artists.

The Sensory Layer and Score for Sensory Arrangements
If there is a particular composition that you have arranged for sensory stimulation or overload, write out the spatial and temporal organization of the onset, peak, and tail of sensory elements. Draw arrows to indicate their movement through the space. Indicate which audience member categories they are designed to engage by drawing a dotted line to connect the shaft of the arrow to the timed location in which those audience engagements occur. If there are attentional patterns that are being controlled with this sensory score, use line weight on the arrow to indicate relative strength of the sensory signal. More salient sensory data gets a wider line weight, while less obvious sensory data is signified by a thin line weight or by some designation, such as a color, that makes sense for you.

The Induced Synesthesia Spatial & Temporal Layer
If there are locations for cross-modal experiences, indicate these on this layer by dividing the space into zones for the crossed experiences. Inside the zones, annotate a list of the linked sensory data, the change in the data (e.g., do the lights and volume ramp up at the same time, and at what specific time and rate of increase?), and the expected effect. This layer might be compared with the sensory, ambient, emotional, and attention layers to identify additional and unanticipated crossing effects.

Information That Does Not Plot Spatially
Some layers of information might not represent well spatially. For instance, although some of it can be represented spatially, not all of the information that deals with time and event sequencing has a spatial component. The non-spatial time information can fit into a multi-channel event sequencing chart similar to Figure 19 to map the timing of events, overlap of events, onset of stimuli, ambient changes, and other temporal non-spatially measured elements.

On a temporal map (a map of the event schedules and places) of the planned experience, annotate the planned use of cognitive mechanisms (conceptual blends, conceptual metaphors, image schemas, viewpoint oscillation, figure-ground organization, attention-structuring events, onset peak and tail for emotional triggers) to organize how different mechanisms combine with the other elements of the experience to create the perceptual relevance of the experience.

Archival Meta-Data Sheet
Metadata helps make sense out of complex datasets. If someone wants to search through an archive to find works that feature specific elements, it is the quality of the metadata associated with the work that ensures that it is found. It also determines the rank a work receives in the search results. Metadata helps when comparing multiple records within a project by giving structure that aids analytics. As a major element of conservation practices in all disciplines, metadata will be specifically useful for the conservation of complex systems and experiences, such as forms of new media art, installation art, or immersive performance art. Like the quote from Michael Mansfield at the beginning of this chapter suggests, documenting time-based works and other new media works is more like documenting biological systems that constantly evolve. These evolving systems provide the basis for developing a list of metadata that might be appropriate for the documentation of your project.

8

Making This Work in a Museum or Gallery

To close, it is useful to think about how the framework in this book applies to the business of art and to suggest future jumping-off points with respect to experimentation. This chapter lays out a model for translating the framework into a museum or gallery setting from the perspective of curatorial staff and other interested parties, such as departments of education, communications, and marketing, and also archivists who each have distinct yet complimentary goals when it comes to educating the public and creating research on collections and shows. Events and interpretive processes can become tools for building research about items in a collection in institutions with permanent collections. The gallery can become a temporary laboratory that sets up an experience of something (whether it is art or some sensory/social relationship) and collects results that can be tied into scholarship, marketing, and other goals of the institution.

Perception Lab

Imagine setting up a sort of "perception lab" in the gallery space to highlight works from the permanent collection in a way that shapes how viewers respond to the works. Viewers engage with the exhibit, but the exhibit is designed as a psychology experiment that looks at how selected features of cognition are at work in viewers as they view individual works in curated exhibits. The aim of this lab-experiment setting is twofold: *educational*—to help viewers experience visual arts from new perspectives—and *curatorial*—to advance collection scholarship through empirical research on human processing of the cognitive elements present in the selected works.

Shaping Visitor Experience

In the pursuit of helping people to have memorable experiences that influence how they see the world in everyday life, one might consider designing enhanced visitor experiences by coupling experimentation, cognitive science, and selected works from the permanent collection. By bringing together these three elements in an exhibit, the exhibit itself can tap into viewer perception in a way that collects patron insight into how viewers participate with art at various levels (such as at the level of individual works, exhibit level, overall museum experience, etc.). Patron insights can be captured along some focused element of perception, such as the perception of time, space, event-structure, viewpoint, agency, objectivity, etc.

This Automatically Helps You Reach Your Institutional Goals

In thinking about how to situate this kind of project in a museum context, it is helpful to start with a set of end goals that might reflect the goals of your institution and then work with these goals as requirements for designing the exhibit.

Curatorial Department Goals

— research the collection
— design substantive exhibits
— present current themes in art practice & scholarship

Education Department Goals

— educate public about collection
— educate public about viewing art in general

— create relevancy

Communications Department Goals

— augment marketing reach
— create repeat visitors
— enrich visibility via community outreach
— reach targeted audience

Museum Goal

— lead the industry in new directions in interpretation and patron-engagement
— establish or maintain reputation as a center of excellence in scholarship

The Visitor Experience
Visitors ("subjects") enter the museum and are directed to a specific gallery, screened, and given a set of perceptual priming activities that both help to hide the research intent of the exhibit experience and also shape the way that they view the experience of the exhibit. The visitors enter the exhibit having received a specific priming (a sort of trained bias) and they engage with each selected work. As they engage with the works, visitors are tasked with making a decision or providing a response that can be captured with some data collection mechanism (perhaps a reaction-timed button-press task, a written response, a multiple choice questionnaire on a mobile app, etc.). Visitors undergo a debriefing interview, and are then possibly put with groups of people who experienced the exhibit with a different sort of priming to get them talking about it. Visitors leave the exhibit and enter the main museum. The data collected from the exhibit is analyzed and then formatted as a numbered edition booklet-mailer which is sent to visitors with an invitation to return to the museum for a special event and presentation of the full report. Depending on available budget and technology support, it might be interesting to determine a more immediate analysis available to visitors that compares results within and across samples of experiment subjects.

Experiment Design
With respect to experimental design, there are several directions you can go:

1. Look at the patron processing of art and see how placards facilitate or interfere with reaction time judgements about the accuracy of the way that the scene is described on the placard. *Most difficult.
2. Prime patrons to look at an exhibit from a certain perspective and collect responses to see how experiences varied. *Moderately difficult.
3. Randomize groups and give them different tours of the same content and get them talking about it in mixed groups afterwards. *Least difficult.
4. Take a hypothesis from a published experimental study in cognitive science, but modify it to use actual art as prompts. *Difficult, but possible.

Your setting and ability to collaborate with cognitive scientists determine the viability of these options, and by no means is this an exhaustive list. Most institutional settings should consider Designs #2 and #3. Scientists know that most experiments fail and that failure is good in the lab because it teaches the scientist something new. However, with respect to public programming in an educational institution like a museum, it is probably better to not risk the failure of an "experiment" and instead focus on building an experience of art that can be informed by science. Instead of trying to break new ground in science, use this experimental experience to break new ground in museum programming and collection scholarship without turning your art museum into a science museum. Again, Designs #2 and #3 will set you up to fulfill these programming, scholarship, and visitor experience-focused goals in a manageable and fruitful manner. Feel free to contact me (visit RyanDewey.org) and

I will connect you with cognitive scientists that can consult with you on your project.

Three Possible Scenarios for Design #2 or Design #3

Subjects arrive and fill out a "plain-English" consent form in a waiting area. Subjects then enter one at a time through a booth. In the booth they are presented with four images, one at a time, each with a linguistic prompt describing the image. They answer a priming question and are measured for reaction time or some other feature. They exit on the other side of the booth, enter the exhibit, and view the exhibit in a predefined order. The exhibit program would use only one of the following three scenarios:

A. Subject-patrons each have a booklet that they are to open as they encounter a new work. The booklet provides a prompt in the form of a placard description that shapes how the work is viewed. There is also a question for each work and the subject must record some kind of response in the booklet: produce a simple drawing, fill in the blank, choose the best description, write a one-sentence description, etc.—anything that generates feedback about how the priming activity is or is not shaping their experience and in a format that is easy to code for analysis. At the end of the exhibit, booklets are collected, subjects are interviewed with some simple debriefing questions, and the interviews are recorded for transcription, coding, and/or analysis.
B. Subject-patrons come to the exhibit, receiving priming similar to Scenario A, but instead of booklets, they have an audio tour or an app that guides them through the exhibit. There are three different versions of the tour and/or app, each one presenting a different perspective on the exhibit. At the end, subject-patrons are debriefed about the exhibit in short interviews that are recorded for transcription, coding, and/or analysis.
C. The exhibit is divided into two sections: *a* & *b*. Subject-patrons are divided into two groups: Group 1 is primed to see section *a* from the vantage point of feature *x* and section *b* from the vantage point of feature *y*. Create the opposite experience for Group 2 by priming them to see section *a* from the vantage point of feature *y* and section *b* from the vantage point of feature *x*. At the end of the exhibit, viewers are debriefed as individuals and then have a conversation about the exhibit with someone who experienced the opposite priming, and this conversation is recorded for transcription, coding, and/or analysis.

Feedback to Patrons

It would be nice to provide some feedback to the viewers/patrons about how they performed in the experiment, and/or the overall results of the experimental exhibit. This could be a report mailer sent to individuals, possibly a special follow-up event that brings some viewers back to the museum to see a presentation of the report, or there could be some social media component (such as using Facebook or Twitter to collect new followers, open communications, and develop marketing lists). In addition, visitors might be presented with real-time feedback if technology resources are available (including information technology personnel with professional-grade analytics).

Drawbacks to the Lab Experience Format

A modern design consideration of many museums is to enable visitors to drive their encounter with the art in order to fit with their browsing pattern, but experimental settings (like laboratories and experiment protocols) typically follow pre-determined progressions rather than browsing patterns. If an experiment protocol is built on an ordered tour that is outside of the control of the visitor, then the exhibit might feel wooden and forced instead of hooking visitors and allowing their own interest to guide them through the exhibit. In other words, the data recording method might not fit with the average visitor behavior (browsing mode). To work around this, consider how your data collection mechanisms can better fit with viewer-centered wayfinding behaviors in the exhibit space. Perhaps a technology solution (like a purpose-designed app or a development patch/modification to your existing app) would enable you to have a scalable data collection method for a visitor-driven experience.

Situating the Exhibit with Other Museum Programming

Various questions need to be answered in order to fit this exhibit into museum programming, such as: What will be exhibiting a year from now? What kinds of programs are planned? What is scheduled for this gallery? Are any visiting curators, researchers, artists, or other residencies planned for that year? How can we coordinate this exhibit with ongoing educational programs? How will this fit within the allotted programming budget?

Fit and Potential Impact

With a novel approach to experiencing art, an experimental exhibit like the Perception Lab may fit well within the intent of your museum or within a specific gallery in your museum. This kind of exhibit will help to transform visitor experiences of the art museum in general, propel visitors into the main museum with new perspectives, equip visitors with new investigative methods to give visitors a clearer understanding of interpretive concepts, and engage viewers by bringing them to art through unique perception-based stories. If this fits with the aims of your museum, this type of exhibit also enables you to establish leadership in interpretive efforts for new museum patronage and may enable you to continue to push the boundaries of museum programming in an increasingly inventive industry.

Outcomes: How it Reaches your Institutional Goals

Curatorial Department Outcomes

- **Research the Collection**: One result from this exhibit would be a body of authentic visceral responses to specific art from the public that will be analyzed and recorded in archives. This could possibly be written up and published. Another result would be a catalogue of specifications for the selected works as seen through the lens of cognitive science.
- **Design Substantive Exhibits**: Curatorial goals aim for substantive design. The approaches to experience design laid out in this book are fairly novel as approaches to exhibit design and are novel enough to act as a guide in defining a new approach to public participation with exhibits that satisfy the requirement for substantive design.
- **Present Current Themes in Art Practice & Scholarship**: This exhibit becomes a participatory work of performance art as viewers embody subjects in an experimental setting. This also brings different research communities together over important topics, such as the role of art in life, and also of perception in daily life.

Education Department Outcomes

- **Educate the Public about the Collection**: The public sees a selection of works from the collection unified with an educational narrative that is unique and engaging. Subsequent communication to the public continues to educate about the collection after the initial visit to the gallery.
- **Educate the Public about Viewing Art in General**: The narratives and didactics build a framework for viewing art from specific perspectives that can then be applied to other viewing experiences.
- **Create Relevancy**: This framework for viewing art from specific perspectives is cognitively relevant because it is perceptually relevant. people see the visual arts differently and with new eyes. This exhibit also educates the public about the importance of cognition and brain/mind research through encounters with art.

Communications Department Outcomes

- **Augment Marketing Reach**: Visitors get feedback from the exhibit by joining a mailing list, and visitor responses can be coded to create member metrics like preferences and behaviors for tailoring marketing in the future.
- **Create Repeat Visitors**: The follow-up events draw repeat visitors to the museum, and the ones that do return come

Visitor Phases	Staffing Needs
Visitor decides to come to your museum	ongoing marketing, web editing & social media visibility, exhibit specific push
Visitor decides to enter the lab gallery	exhibit specific push, exterior & interior signage, personal engagement, entrance staffing
Visitor becomes willing to engage with an interactive experimental experience	screen visitor, obtain consent,
Visitor interacts with the experience	lab gallery interns, guards, possibly docents
Visitor is debriefed about the experience	administer questionnaire, audio/visual recording, conversation facilitators/moderators
Visitor exits the lab gallery	exit staffing
Visitor enters main museum and experiences museum with new knowledge gained in the lab gallery	regular ongoing staffing, possible interpretive tours or waypoints throughout museum that relate to exhibit
Visitor exits the museum by walking past the lab gallery	engage visitors with a handout or save the date card
Visitor tells others about the experience	copywriting, web editing & social media visibility for exhibit info, art direction, postcard production
Visitor going about their everyday life	transcribing data, coding, analysis, developing visualizations & graphics, finding a story in data
Visitor receives mailer feedback	marketing, art direction, production of booklet, mailing list database management
Visitor returns for the special follow-up experience	event planning, facilities, speakers & entertainment
Visitor returns to your museum for their next visit	ongoing regular marketing push

Table 11. Visitor Phases and Staffing Needs.

back with a renewed purpose for visiting the museum as participant-stakeholder.

- **Enrich Visibility via Community Outreach**: In coordination with an education department, programming around this exhibit can reach out to children (future museum patrons) and can strengthen relationships with educational institutions.
- **Reach Targeted Audience**: Metrics captured in visitor responses might suggest potential individuals to pursue for museum membership. Metrics also provide insight into which program offerings are most relevant to particular list members.

General Museum Outcomes

- **Lead the Industry in New Directions in Interpretation and Patron-Engagement**: This exhibit continues in the spirit of leadership, perhaps capturing the eyes of the museum world, and takes a new approach to helping patrons understand the significance of art.

Possible First Steps

Use these possible first steps as a jumping off point, bring these steps to your planning table and rip them apart and build from them. Here is one way to approach this project:

- Organize a working group consisting of program personnel to discuss the fit with your institution.
- Work with a cognitive scientist to identify suitable works in the existing collection. Begin by looking for works that strongly feature ambiguity along these dimensions: *time, space, viewpoint, agency, place, activity/event,* and *object*. You need to figure out early on what you have to work with so you can find a story in the collection.
- Send images of works to everyone in the working group, schedule a meeting to discuss which works might fit which experiment/experience structure. Decide on two approaches so that you have a backup plan if the preferred project fails or seems unrealistic.
- Use the various visitor phases specified in Table 11 as a skeleton and build the experience on that armature.
- Use the visitor phases to outline staffing tasks and role/job descriptions for personnel.
- Begin by using Design #2, Scenario B. Test it out with staff and friends early and on a small scale. If 2.B works for you in your test, extend it to a fuller scale. If it succeeds with the public, create a variation that evolves the design. Only move on to a more difficult design after trying 2.B in your local context.

Bibliography

Antovic, Mihailo, Austin Bennett, and Mark Turner. 2013. "Running in Circles or Moving Along Lines: Conceptualization of Musical Elements in Sighted and Blind Children." *Musicae Scientiae* 17, no. 2: 229–45; https://doi.org/10.1177/1029864913481470.

Auvray, Malika and Mirko Farina. 2017. "Patrolling the Boundaries of Synaesthesia: A Critical Appraisal of Transient and Artificially-Acquired Forms of Synaesthetic Experiences." In *Sensory Blending: New Essays on Synaesthesia,* ed. Ophelia Deroy, 248–73. Oxford: Oxford University Press.

Bang, Molly. 2000. *Picture This: How Pictures Work.* San Francisco: Chronicle Books.

Battaglia, Fortunato, Sarah H. Lisanby, and David Freedberg. 2011. "Corticomotor Excitability During Observation and Imagination of a Work of Art." *Frontiers in Human Neuroscience* 5, art. 79: 1–6; https://doi.org/10.3389/fnhum.2011.00079.

Bell, E.A., L.S. Roe, and Barbara J. Rolls. 2003. "Sensory-Specific Satiety is Affected more by Volume than by Energy Content of a Liquid Food." *Physiology & Behavior* 78, nos. 4–5: 593–600; https://doi.org/10.1016/S0031-9384(03)00055-6.

Bergen, Benjamin and Kathryn Wheeler. 2010. "Grammatical Aspect and Mental Simulation." *Brain & Language* 112, no. 3:150–58; https://doi.org/10.1016/j.bandl.2009.07.002.

Booker, Christopher. 2004. *The Seven Basic Plots: Why We Tell Stories.* London: Continuum.

Burke, Kenneth. 1969. *A Grammar of Motives.* Berkeley: University of California Press.

Brown, Caleb and Ryan Dewey. 2014. "Being There: Attention and Anticipation in Film." Conference presentation, 2nd Annual "Screening Scholarship Media Festival," March 2, Annenberg School for Communication, University of Pennsylvania.

Chafe, Wallace. 1994. *Discourse, Consciousness, and Time: The Flow and Displacement of Conscious Experience in Speaking and Writing.* Chicago: University of Chicago Press.

Clark, Herbert H. 1996. *Using Language.* Cambridge: Cambridge University Press.

Cocteau, Jean. 1921. *Cock and Harlequin: Notes Concerning Music,* trans. Rollo H. Meyers. London: Egoist Press. https://archive.org/details/CockAndHarlequin.

Darwin, Charles. 2016. *The Works of Charles Darwin, Vol. 23: The Expression of the Emotions in Man and Animals,* ed. Francis Darwin. London: Routledge.

Dewey, John. 2005. *Art as Experience.* Berkeley: Perigee.

Dewey, Ryan. 2012. *A Sense of Space: Conceptualization in Way-Finding and Navigation.* M.A. Thesis, Case Western Reserve University, http://rave.ohiolink.edu/etdc/view?acc_num=case1339097784.

———. 2014. "Agency and the Multifaceted Stories of Hybrid Places." *MONU Magazine* #20: 78–83.

———. 2014. "Hacking Remoteness Through Viewpoint and Cognition." *KERB: Journal of Landscape Architecture* 22: 26–33.

Donald, Merlin. 2006. "Art and Cognitive Evolution." In *The Artful Mind: Cognitive Science and the Riddle of Human Creativity,* ed. Mark Turner, 3–20. Oxford: Oxford University Press.

Dunne, Anthony and Fiona Raby. 2013. *Speculative Everything: Design, Fiction, and Social Dreaming.* Cambridge: MIT Press.

Egermann, Hauke, Nathalie Fernando, Lorraine Chuen, and Stephen McAdams. 2014. "Music Induces Universal Emotion-Related Psychophysiological Responses: Comparing Canadian

Listeners to Congolese Pygmies." *Frontiers in Psychology* 5, art. 1341: 1–9; https://doi.org/10.3389/fpsyg.2014.01341.

Evans, Vyvyan. 2009. *How Words Mean: Lexical Concepts, Cognitive Models, and Meaning Construction.* Oxford: Oxford University Press.

Fauconnier, Gilles and Mark Turner. 1998. "Conceptual Integration Networks." *Cognitive Science* 22, no. 2: 133–87; https://doi.org/10.1207/s15516709cog2202_1.

Flusberg, Stephen J. and Lera Boroditsky. 2011. "Are Things that Are Hard to Physically Move Also Hard to Imagine Moving?" *Psychonomic Bulletin & Review* 18, no. 1: 158–64; https://doi.org/10.3758/s13423-010-0024-2.

Freedberg, David. 2006. "Composition and Emotion." In *The Artful Mind,* ed. Mark Turner, 73–89. Oxford: Oxford University Press.

Freedberg, David and Vittorio Gallese. 2007. "Motion, Emotion and Empathy in Esthetic Experience." *Trends in Cognitive Sciences* 11, no. 5: 197–203; https://doi.org/10.1016/j.tics.2007.02.003.

Freud, Sigmund. 2005. *Civilization and Its Discontents* (1929), trans. James A. Strachey. New York: W.W. Norton & Co.

Gibbs Jr., Raymond W. 2000. "Making Good Psychology Out of Blending Theory." *Cognitive Linguistics* 11, nos. 3/4: 347–58; https://dx.doi.org/10.1515/cogl.2001.020.

Glenberg, Arthur M. and Michael P. Kaschak. 2002. "Grounding Language in Action." *Psychonomic Bulletin & Review* 9, no. 3: 558–65; https://doi.org/10.3758/BF03196313.

Hodder, Ian. 2012. *Entangled: An Archaeology of the Relationships Between Humans and Things.* Oxford: Wiley-Blackwell.

Hubbard, Edward M. 2007. "Neurophysiology of Synesthesia." *Current Psychiatry Reports* 9, no. 3: 193–99; https://doi.org/10.1007/s11920-007-0018-6.

Hyman, John. 2010. "Art and Neuroscience." In *Beyond Mimesis and Convention: Representation in Art and Science,* eds. Roman Frigg and Matthew Hunter, 245–61. Dordrecht: Springer.

Ingels, Bjarke and Kai-Uwe Bergmann. 2014. *THE BIG MAZE* (museum exhibit), National Building Museum, Washington, D.C., July 4–September 1, https://www.nbm.org/exhibition/the-big-maze/.

Jones, Caroline A. 2006. *Sensorium: Embodied Experience, Technology, and Contemporary Art.* Cambridge: MIT Press.

Keller, Thomas, Susie Heller, Michael Rulhman, and Deborah Jones. 1999. *The French Laundry Cookbook.* New York: Artisan.

Lakoff, George. 2006. "The Neuroscience of Form in Art." In *The Artful Mind: Cognitive Science and the Riddle of Human Creativity,* ed. Mark Turner, 153–69. Oxford: Oxford University Press.

Lakoff, George and Mark Johnson. 1980. *Metaphors We Live By.* Chicago: University of Chicago Press.

———. 1999. *Philosophy in the Flesh: The Embodied Mind and Its Challenge to Western Thought.* New York: Basic Books.

Lee, Spike S.W. and Norbert Schwarz. 2012. "Bidirectionality, Mediation, and Moderation of Metaphorical Effects: The Embodiment of Social Suspicion and Fishy Smells." *Journal of Personality and Social Psychology* 103, no. 5: 737–49; https://dx.doi.org/10.1037/a0029708.

Lynch, Kevin. A. 1960. *The Image of the City.* Cambridge: MIT Press.

Mansfield, Michael. 2013. "Challenges in the Curation of Time Based Media Art: An Interview with Michael Mansfield" (interview by Jose [Ricky] Padilla). *The Signal* (Library of Congress weblog), April 9, http://blogs.loc.gov/thesignal/2013/04/challenges-in-the-curation-of-time-based-media-art-an-interview-with-michael-mansfield/.

Matlock, Teenie. 2004. "The Conceptual Motivation of Fictive Motion." In *Studies in Linguistic Motivation,* eds. Günter Radden and Klaus-Uwe Panther, 221–48. Berlin: Mouton de Gruyter.

———. 2004. "Fictive Motion as Cognitive Simulation." *Memory & Cognition* 32, no. 8: 1389–1400; https://doi.org/10.3758/BF03206329.

Noë, Alva. 2004. *Action in Perception.* Cambridge: MIT Press.

Oakley, Todd. 2009. *From Attention to Meaning: Explorations in Semiotics, Linguistics, and Rhetoric.* Bern: Peter Lang.

Ramachandran, V.S. and William Hirstein. 1999. "The Science of Art: A Neurological Theory of Aesthetic Experience." *Journal of Consciousness Studies* 6, nos. 6/7: 15–51.

Rainer, Yvonne. 2006. "Labanotation and Trio A." In *Sensorium: Embodied Experience, Technology, and Contemporary Art,* ed. Caroline A Jones, 163–66. Cambridge: MIT Press.

Rittel, Horst W.J. and Melvin M. Webber. 1973. "Dilemmas in a General Theory of Planning." *Policy Sciences* 4, no. 2: 155–69; https://doi.org/10.1007/BF01405730.

Robinson, David L. 2008. "Brain Function, Emotional Experience and Personality." *Netherlands Journal of Psychology* 64, no. 4: 152–68; https://doi.org/10.1007/BF03076418.

Rolls, Edmund T. and J.H. Rolls. 1997. "Olfactory Sensory-Specific Satiety in Humans." *Physiology & Behavior* 61, no. 3: 461–73; https://doi.org/10.1016/S0031-9384(96)00464-7.

Rolls, Edmund T., Barbara J. Rolls, and E.A Rowe. 1983. "Sensory-Specific and Motivation-Specific Satiety for the Sight and Taste of Food and Water in Man." *Physiology & Behavior* 30, no. 2: 185–92; https://doi.org/10.1016/0031-9384(83)90003-3.

Sato, Manami, Amy J. Schafer, and Benjamin K. Bergen. 2013. "One Word at a Time: Mental Representations of Object Shape Change Incrementally During Sentence Processing." *Language and Cognition* 5, no. 4: 345–73; https://doi.org/10.1515/langcog-2013-0022.

Sbriscia-Fioretti, Beatrice, Cristina Berchio, David Freedberg, Vittorio Gallese, and Maria Alessandra Umiltà. 2013. "ERP Modulation During Observation of Abstract Paintings by Franz Kline." *PLoS ONE* 8, no. 10: e75241; https://doi.org/10.1371/journal.pone.0075241.

Semken, Steve, Jeff Dodick, Orna Ben-David, Monica Pineda, Nievita Bueno Watts, and Karl Karlstrom. 2009. "Timeline and Time Scale Cognition Experiments for a Geological Interpretative Exhibit at Grand Canyon." Conference presentation, Proceedings of the National Association for Research in Science Teaching, Garden Grove, California. http://semken.asu.edu/pubs/semken09_tatex.pdf.

Schwartzman, Madeline. 2011. *Seeing Yourself Sensing: Redefining Human Perception.* London: Black Dog Publishing.

Silbert, Lauren, Jennifer Silbert, Suzanne Dikker, Matthias Oostrik, and Oliver Hess. 2012. *Compatibility Racer: A Project Examining the Art of Interactive Neuroscience* (website), http://compatibilityracer.blogspot.com/.

Slepian, Michael L. and Nalini Ambady. 2014. "Simulating Sensorimotor Metaphors: Novel Metaphors Influence Sensory Judgments." *Cognition* 130, no. 3: 309–14; http://dx.doi.org/10.1016/j.cognition.2013.11.006.

Spradley, James P. 2016. *Participant Observation.* Long Grove: Waveland Press.

Stern, Nathaniel. 2013. *Interactive Art and Embodiment: The Implicit Body as Performance.* Canterbury: Gylphi Limited.

Talmy, Leonard. 2008. "Aspects of Attention in Language." In *Handbook of Cognitive Linguistics and Second Language Acquisition,* eds. Peter Robinson and Nick C. Ellis, 27–38. New York: Routledge.

Talmy, Leonard. 2001. *Toward a Cognitive Semantics, Vol. 1: Concept Structuring Systems.* Cambridge: MIT Press.

Thiis-Evensen, Thomas. 1987. *Archetypes in Architecture.* Oslo: Norwegian University Press.

Tobias, Ronald B. 1993. *20 Master Plots: And How to Build Them.* Cincinnati: Writer's Digest Books.

Tobin, Vera and Todd Oakley. 2012. "Attention, Blending, and Suspense in Classic and Experimental Film." In *Blending and the Study of Narrative: Approaches and Applications,* eds. Ralf Schneider and Marcus Hartner, 57–84. Berlin: Walter de Gruyter.

Tomasello, Michael. 2003. *Constructing a Language: A Usage-Based Theory of Language Acquisition.* Cambridge: Harvard University Press.

Tversky, Barbara. 2011. "Spatial Thought, Social Thought." In *Spatial Schemas in Social Thought,* eds. Thomas W. Schubert and Anne Maass, 17–38. Berlin: De Gruyter Mouton.

Umiltà, Maria Alessandra, Cristina Berchio, Mariateresa Sestito, David Freedberg, and Vittorio Gallese. 2012. "Abstract Art and

Cortical Motor Activation: An EEG Study." *Frontiers in Human Neuroscience* 6, art. 311: 1–9; https://dx.doi.org/10.3389%2Ffnhum.2012.00311.

brainstorm books

"The local mechanisms of mind...are not all in the head.
Cognition leaks out into body and world."

— Andy Clark, *Supersizing the Mind*

Current developments in psychoanalysis, psychology, philosophy, and cognitive and neuroscience confirm the profound importance of expression and interpretation in forming the mind's re-workings of its intersubjective, historical and planetary environments. Brainstorm Books seeks to publish cross-disciplinary work on the becomings of the extended and enactivist mind, especially as afforded by semiotic experience. Attending to the centrality of expression and impression to living process and to the ecologically-embedded situatedness of mind is at the heart of our enterprise. We seek to cultivate and curate writing that attends to the ways in which art and aesthetics are bound to, and enhance, our bodily, affective, cognitive, developmental, intersubjective, and transpersonal practices.

Brainstorm Books is an imprint of the "Literature and the Mind" group at the University of California, Santa Barbara, a research and teaching concentration hosted within the Department of English and supported by affiliated faculty in Comparative Literature, Religious Studies, History, the Life Sciences, Psychology, Cognitive Science, and the Arts.

http://mind.english.ucsb.edu/brainstorm-books/

Brainstorm Books

Made in the USA
Columbia, SC
08 December 2018